LEARN MATH FAST SYSTEM
VOLUME III
FOURTH EDITION

By J. K. Mergens

Learn Math Fast System Volume III, Fourth Edition
Published by MGM Publishing
Copyright © 2011 © 2019 ©2021 Registration Number TX 7-316-060
ISBN: 978-1519597441
www.LearnMathFastBooks.com

United States Coin Image from the United States Mint

CONTENTS

Introduction 5

Intro to Pre-Algebra 7
    Lesson 1: What is x? 7
    Lesson 1a: Solve for x with Fractions 14
    Lesson 2: Solve for x with multiplication 18
    Lesson 2a: Solve for x with Negative Numbers 22
    Lesson 3: Solve for x with Multiplication 27
    Lesson 4: Solving for Negative X 30
    Lesson 5: Solving for x with Division 35
    Lesson 6: What is y? 39
    Lesson 7a: Ratios 43
    Lesson 7b: Working with Ratios 49
    Lesson 8a: Proportions 54
    Lesson 8b: Unit Conversion 58
    Lesson 9: Exponents^2 64
    Lesson 10: More Exponents^3 66
    Lesson 11: Square Root 69
    Chapter 1 Review Test 72

Solving Algebraic Expressions 73
    Lesson 12: Terms, Expressions and Equations 73
    Lesson 13: Combining Like Terms 76
    Lesson 14: Multiplying Terms 82
    Lesson 15: Math Inside of Parentheses 86
    Lesson 16: The Second Operation 89
    Lesson 17: P.E.M.D.A.S 96
    Lesson 18: Distributive Property of Multiplication 102
    Lesson 19: Solving Algebraic Equations 107
    Lesson 19a: Solving Expressions with the Distributive Property of Multiplication 112
    Chapter 2 Review Test 117

Slopes 119
    Lesson 20: Graphs 119
    Lesson 21: Linear Equations 127
    Lesson 22: Slope of a Line 134
    Lesson 23: The Y-Intercept 143

Lesson 24: Creating a Linear Equation 148
Lesson 25: Slopes Review 152
Chapter 3 Review Test 154
Final Test 156
ANSWERS 158

INTRODUCTION

Welcome to Volume III of the *Learn Math Fast System*. For best results, please read Volumes I & II before reading this book. But, if you already know how to solve fractions and work with negative numbers, then you are ready for this book.

There is a worksheet at the end of each lesson. The answers to each worksheet are in the back of the book. Check your answers after each worksheet to make sure you are getting the right answers. If you get stuck, use the answer key to help you solve the problem. If you get stuck more than once on a worksheet, read the lesson again and then start over.

When you are finished with this book, you are ready for Volume IV of the *Learn Math Fast System*. Volume IV covers up to 8th grade geometry.

If you have any questions or comments, please contact us via our website www.LearnMathFastBooks.com

CHAPTER 1
INTRO TO PRE-ALGEBRA
LESSON 1: WHAT IS X?

Pre-algebra can be quite simple, once you get past the big mystery of "x." Many people get lost in math as soon as the letter "x" gets involved. Let me solve the mystery for you. The letter "x" is just a question mark! For example, look at this simple equation, 2 + 3 = 5. Now look at the algebra equation below. Can you guess how much "x" is?

$$2 + x = 5$$

The answer is 3. To be exact, x = 3. Can you solve this problem?

$$x + 3 = 5$$

How much is x now? x = 2. Math people don't usually say, "How much is x?" Instead, they say, "Solve for x."

Since you know 2 + 3 = 5, it is easy to solve for x in the problems above, but it's not always that simple. Sometimes, you have to use algebra to get the answer. Here is how you would figure out the problem above with algebra, if you couldn't solve it in your head. It's easy to do, if you remember the steps.

First, the big trick in algebra is to "GET X BY ITSELF!" That means, we want all the numbers to be on one side of the equal sign and just x on the other side, all by itself. That way we are left with x = *something*, and that *something* is your answer. Look at the algebraic *equation* below. An *equation* is a math statement that says, "This equals that."

$$x + 3 = 5$$

To start, we need to *get x by itself*. That means we need to move the 3 over to the other side of the equal sign. That's what I mean by having all the numbers on one side of the equal sign and the x on the other side; all by itself.

In algebra, the only way to move that 3 over to the other side is by doing the *opposite operation*. *Operation* is a fancy word for plus, minus, multiply, or divide. The word *opposite* means to undo something. For example, the opposite of "off" is "on." They undo each other. The opposite of "up" is "down," they undo each other, too.

The opposite of addition is subtraction. They are *opposite operations*. For example, let's say your sister *added* onions to your pizza. You want to undo that, so you *subtract* the onions from the pizza. You just got rid of something that was added by doing the *opposite operation*. Let's get back to using algebra to solve the simple problem below.
$$x + 3 = 5$$

We have to do the *opposite operation*, to move + 3 to the other side. The opposite of plus 3 is minus 3, so let's minus 3 from this side of the equal sign.

$$\begin{array}{r} x + 3 = 5 \\ -3 \end{array}$$

That way, x will be all by itself. But, if we minus 3 from that side of the equal sign, then we HAVE TO minus 3 from the other side, too. Think of the equal sign as the center point of a scale. You must keep each side balanced. If you subtract 3 from one side, you must do that to the other side, so you don't tip the scale.

If you minus 3 from here,

you'll tip the scale.

Put the "minus 3" on the other side.   x = 5 - 3

Now both sides are balanced. So far, all we have done is subtracted 3, to get x by itself and then put that -3 on the other side.

Here is how you write the math to show subtracting 3 from both sides. This math is solved vertically ↓ from top to bottom, one section at a time. Bring the x straight down because it didn't change.

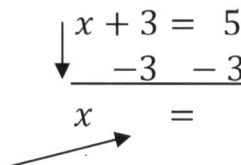

Next in line is +3 − 3. That is the same thing as 3 − 3, which is 0 or nothing, so it goes away. Bring down the equal sign and subtract 5 − 3.

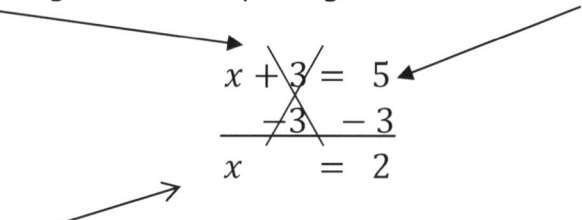

The answer is x = 2. That was a long explanation for a small problem, so let's make sure you get the main points.

The first step is to get x all by itself. To do that you need to do the opposite operation. That's how you move a number away from x. Next, whatever you do on one side of the equal sign, you MUST do on the other side of the equal sign, so your scale doesn't tip. Then just do the math.

Look at this next one. We will slowly go over each step again.

$$x - 10 = 20$$

Step 1:

Get x by itself. We need to move the -10 to the other side to get x by itself. This is minus 10, so we need to do the *opposite* to make it disappear. The opposite is plus 10. The math is -10 + 10 = 0. It disappears!

Step 2:

Whatever you do to one side of the equal sign you MUST do to the other side, so you don't tip the scale. Add 10 to both sides and then solve the math vertically.

$$x - 10 = 20$$
$$+10 +10$$

Bring the x straight down; it is unchanged. Do the math -10 + 10 = 0 and 20 + 10 = 30.

$$x - \cancel{10} = 20$$
$$\cancel{+10} +10$$
$$x = 30$$

Look back at the original equation; x – 10 = 20. Can you see why x = 30? Put "30" in place of the x. You get 30 – 10 = 20. That's why x = 30.

Here's a short cut. To make it even easier, you don't have to write down the opposite math; we just ended up crossing it out anyway. Instead, **just swing the number over to the other side and change the sign to the opposite sign**. I'll show you what I mean. We will do those last two problems again with this easy short cut way.

x + 3 = 5	Swing the 3 over and change the sign to a minus.
x = 5 - 3	Do the math; 5 - 3 = 2.
x = 2	That is much faster.

Here is the second problem, a little easier.

x - 10 = 20	Move the 10 to the other side and change the sign to +.
x = 20 + 10	Do the math; 20 + 10.
x = 30	You just solved for x. The answer is x = 30.

You can use which ever method is easier for you. Write down the opposite math or just swing the number to the other side and change the + or – sign. Take a look at this next example.

$$x - 10 = -3$$

Start by swinging that 10 over to the other side, but first change the sign to make it +10. You may be wondering, which is the correct way to write that?

Like this:

$$x = -3 + 10$$

Or like this:

$$x = 10 - 3$$

Surprisingly, you will end up with the same answer either way; they both equal 7. The important part is to make sure you are transferring the signs correctly. Let's try another one.

$$12 + x = -4$$

This time we need to move this 12. What is it, positive or negative 12? Since there is no sign in front, it must be positive. So, if it is a positive 12, then we need to do the opposite to get rid of it. I will subtract 12 from both sides of the equation.

$$\begin{aligned} 12 + x &= -4 \\ -12 & -12 \end{aligned}$$

So, let's see, how should we write that? Negative 4 minus 12 or negative 12 minus 4?

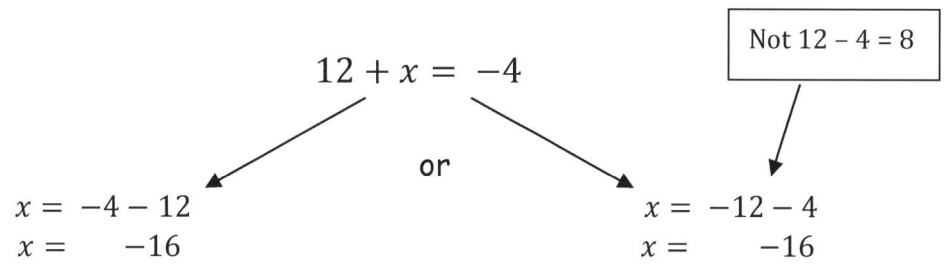

Either way you get the same answer. Just make sure you don't accidentally write 12 – 4; it is NEGATIVE 12 MINUS 4.

Try some on your own. As you complete the next worksheet, remember the two steps.

Step 1: Use an opposite operation to get x by itself.
Step 2: Whatever you do to one side, you must do to the other side.

Name: _____   Date: _____

WORKSHEET 3-1

Solve for x.

1. $4 + x = 24$
2. $x + 14 = 21$
3. $2 + x = 12$

4. $x + 72 = 172$
5. $x - 33 = 54$
6. $9 + x = 62$

7. $4 + x = 0$
8. $x + 25 = 100$
9. $x - 8 = 176$

10. $10 + x = 310$
11. $x + 38 = 44$
12. $x - 8 = 34$

13. $x - 8 = 5$
14. $x - 14 = -2$
15. $9 + x = -1$

16. $11 + x = -7$
17. $4 + x = 2$
18. $x - 3 = 27$

19. $x + 36 = 36$
20. $4 + x = -26$
21. $-33 = 7 + x$

22. Eric had 57 baseball cards. He gave his brother a small handful of them. Now he only has 43 cards left. How many cards did Eric give to his brother?

$$43 + x = 57$$

23. Mike has 9 gallons of paint. He needs a total of 17 gallons to paint the house. How many more gallons does he need?

$$9 + x = 17$$

LESSON 1A: SOLVE FOR X WITH FRACTIONS

Now that you've had a chance to work with "x" a little bit, I'm going to make the problems a little more interesting. The math is the same, but I will throw in some fractions, mixed numbers and negative numbers.

Again, the math is the same. You still need to get x by itself, and you still need to make sure you do the same thing to both sides of the equation. Let's try one together.

The first thing we need to do is get that x by itself.

$$\frac{1}{3} + x = -\frac{2}{3}$$

To do that, we need to subtract that $\frac{1}{3}$ from the left-hand side of the equal sign. I'll subtract it from that side, and of course, whatever we do to one side of the equal sign, we must do to the other side, too. I will subtract $\frac{1}{3}$ from both sides to get x by itself.

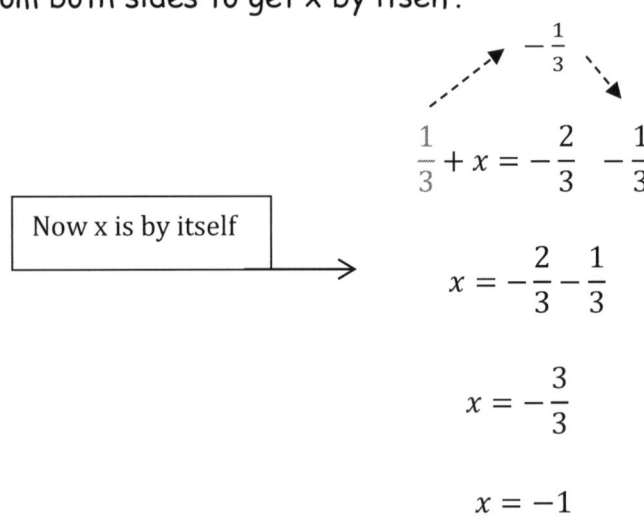

Now x is by itself

$$\frac{1}{3} + x = -\frac{2}{3} \quad -\frac{1}{3}$$

$$x = -\frac{2}{3} - \frac{1}{3}$$

$$x = -\frac{3}{3}$$

$$x = -1$$

Do you see how the math is the same? It's just a little more complicated. Now I'll throw in a negative mixed number.

$$-3\frac{5}{8} + x = \frac{1}{2}$$

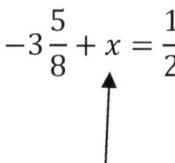

Where do we start? We have to get that x by itself. The only number on that side of the equal sign is a negative number, so let's do the opposite math and ADD $3\frac{5}{8}$. And whatever we do to one side, we have to do to the other side, too. After adding $3\frac{5}{8}$ to both sides, I'm left with this:

$$x = \frac{1}{2} + 3\frac{5}{8}$$

In order to add these together, I will need to get a common denominator. That's no problem; I'll turn $\frac{1}{2}$ into $\frac{4}{8}$. Now it's easy to add them together.

$$x = \frac{4}{8} + 3\frac{5}{8}$$

$$x = 3\frac{9}{8}$$

$$x = 4\frac{1}{8}$$

Now you try one. This time we will use decimal numbers.

$$.85 + x = -1.35$$

Do you know what to do? That's right, you need to get rid of the .85. Is it positive or negative? It is positive. OK, then let's subtract it from both sides or just swing it over to the other side and change the sign.

$$.85 + x = -1.35 - .85$$

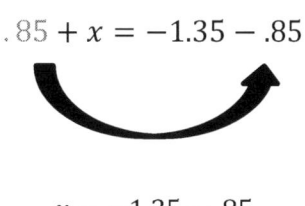

$$x = -1.35 - .85$$

Now all we have to do is the math. So, let's see...a negative number minus a positive number...that's like owing $1.35 and then someone comes along and takes another 85 cents from you. Now you have even less money. I'll add the two numbers together and call it negative.

$$x = -2.20$$

Give it a try for yourself on the next worksheet. If you are having difficulties working with negative numbers or fractions, read volume II of the *Learn Math Fast System*.

Name: _____   Date: _____

WORKSHEET 3-1a

Solve for x.

1. -3 + x = -12   2. x - 14 = -28   3. 12 + x = 4

4. 10 + x = 5   5. x + 8 = -48   6. 62 = x + 9

7. -44 = x - 18   8. x + 25 = -100   9. x - 10 = -17

10. x + 5 = -55   11. x + 7 = -12   12. 8 + x = -20

13. x - 27 = -15   14. $-\frac{1}{2} + x = -.5$ 15. -.75 + x = .25

16. x + .3 = -5.3   17. $\frac{1}{2} + x = 5$ 18. $x - \frac{1}{4} = -6\frac{3}{4}$

19. Sarah is trying to break the record for doing the most one-handed cartwheels on a balance beam without falling. Right now, the record is 71, so she needs to get to 72 to break the record. She has done 15 cartwheels, so far. How many more does she need to do to break the record? Use algebra to solve for x in the equation below.

$$15 + x = 72$$

20. Robin needs to keep track of the water level at Lake Welch. At the end of the summer, the water level was low. It measured 30 inches below the desired level. After a week of rain, the water level rose and is now only 16 inches below the desired level. How many inches did it rain? Use algebra to solve for x.

$$-30 + x = -16$$

LESSON 2: SOLVE FOR X WITH MULTIPLICATION

If you had problems getting the correct answers on that last worksheet, read that lesson again. It will be easier the second time around.

The problem below is saying, "3 times x equals 21." It's kind of confusing to have an x next to a multiplication sign.

$$3 \times x = 21$$

To make it easier to read, we can replace the multiplication sign with a dot.

$$3 \cdot x = 21$$

This is also read as, "3 times x equals 21." To make it even easier, we can forget about the dot and just squish the 3 and x together.

$3 \cdot x = 21$	→ [3 x] ← = 21	$3x = 21$
TOSS THE DOT	→ [SQUISH] ←	NEW SLIM LOOK

When a number is next to a letter, such as 3x, it means to multiply. Whether you write it with a dot or next to each other, it means the same thing; multiply.

Do you remember learning that addition is the opposite of subtraction? Now it is time to learn the opposite of multiplication. Can you guess what the opposite of multiplication is? The opposite of multiplication is division. In order to solve for x in this next problem, we need to get rid of that 3, so x is by itself (step 1). The x is *multiplied by 3*, so let's do the opposite - *divide by 3*. That will get x by itself.

$$
\begin{array}{rcl}
3x & = & 21 \\
\div 3 & & \div 3 \\
x & = & 7
\end{array}
$$

(Step 2) Whatever you do to one side of the equal sign you MUST do to the other side, to keep it equal, so divide both sides by 3 to get x by itself.

To find out if we got the right answer, replace the x with our answer. The problem is 3x = 21. We say x = 7, so let's replace the x with our 7.

$$3 \cdot 7 = 21$$

3 x 7 does equal 21, so we have the right answer.

Here is another way to write out the same problem. I know you are familiar with the division symbol that looks like this ÷. But in algebra, they have *another* symbol for division. Here is the same math with the new symbol for division; it is an underline.

$$\frac{3x}{3} = \frac{21}{3}$$

This means $3x \div 3 = 21 \div 3$

$$x = 7$$

They look like fractions, don't they? That's because they are. 21 ÷ 3 is the same thing as $\frac{21}{3}$. If you reduce that down, you will get 7. (If that doesn't make sense, read the previous book in this series).

You could also solve that problem in your head by looking at what the problem is asking.

$$3x = 21$$

Now think to yourself, "3 times *what* equals 21?" 3 x 7 = 21, so x = 7.

Look at these next 4 problems. Can you solve for x? Remember the steps. Step 1: Use the opposite operation to get x by itself. Step 2: Whatever you do to one side, you must do to the other side. Or just simply solve them in your head.

4x = 16   5x = 25   7x = 42   9x = 18

Do you remember the short cut from earlier? The short cut is "swing the number over to the other side and change the sign." This is how to use the shortcut for the four problems on the last page. "Change the sign" means do the opposite operation, so multiplication changes to division.

$\cancel{4}x = 16 \div 4 \qquad \cancel{5}x = 25 \div 5 \qquad \cancel{7}x = 42 \div 7 \qquad \cancel{9}x = 18 \div 9$

Now can you solve for x?

$x = 16 \div 4 \qquad x = 25 \div 5 \qquad x = 42 \div 7 \qquad x = 18 \div 9$

$x = 4 \qquad\qquad x = 5 \qquad\qquad x = 6 \qquad\qquad x = 2$

Practice solving for x on your own by completing the next worksheet.

Name: _____ Date: _____

WORKSHEET 3-2

Solve for x.

1. $4x = 16$
2. $5x = 15$
3. $3x = 21$
4. $9x = 54$

5. $8x = 56$
6. $2x = 12$
7. $8x = 72$
8. $3x = 12$

9. $9x = 63$
10. $4x = 28$
11. $5x = 105$
12. $11x = 176$

13. $10x = 210$
14. $12x = 144$
15. $8x = 24$
16. $6x = 24$

17. $1/2x = 4$
18. $3x = 42$
19. $9x = 81$
20. $11x = 154$

21. $4x = 32$
22. $7x = 42$
23. $6x = 48$
24. $6x = 36$

25. $2x = 100$
26. $4x = 36$
27. $9x = 18$
28. $49 = 7x$

29. $9 = 3x$
30. $27 = 3x$
31. $30 = 6x$
32. $48 = 12x$

33. $11x = 33$
34. $12x = 60$
35. $14x = 56$
36. $100x = -500$

LESSON 2A: SOLVE FOR X WITH NEGATIVE NUMBERS

The problems on the last worksheet were nice and easy, so let's bump it up a little. Look at the problem below. The math is the same as before, but we have to deal with a negative mixed number this time.

$$-2\frac{3}{4}x = -11$$

Can you figure out what to do? Let's start by reading this problem. It is negative two and three fourths, times x, equals negative eleven. The first step is always to get x by itself. The way to do that is by doing the opposite operation. Since x is multiplied by $-2\frac{3}{4}$, we want to DIVIDE by $-2\frac{3}{4}$. That will get x by itself. And what is the second step? Whatever you do to one side, you must do to the other side of the equal sign. Otherwise, the scale would tip. Now our problem looks like this:

$$x = -11 \div -2\frac{3}{4}$$

Notice that I didn't change the sign of the mixed number. We only change the OPERATION that is being used in the equation. The OPERATION being used is multiplication. The opposite of multiplication is division, so I need to DIVIDE by negative $2\frac{3}{4}$.

$$x = -11 \div -2\frac{3}{4}$$

Do you know how to solve this problem? It is a negative number divided by a negative number, so I know the answer will be positive. In order to divide with a mixed number, we need to turn it into an improper fraction. I will do that and then I'll put the eleven over a one, because that's how you write 11 as a fraction.

$$x = -\frac{11}{1} \div -\frac{11}{4}$$

$2 \times 4 + 3 = 11$

If you recall, the way to divide a fraction is to flip (get the reciprocal) the second fraction and then multiply.

$$x = -\frac{11}{1} \times -\frac{4}{11}$$

When you multiply fractions, you can cross cancel to make the math easier.

$$x = -\frac{\cancel{11}^{1}}{1} \times -\frac{4}{\cancel{11}_{1}}$$

$$x = 4$$

Let's see if I got the right answer. I will bring back the original problem.

$$-2\frac{3}{4} x = -11$$

Now I will replace the x with the answer I got (4) and then I'll do the math to see if it really does equal -11.

$$-2\frac{3}{4} \times 4 =$$

So, let's see...a negative number times a positive number? Since the signs are different, I know the answer will be negative. I have turned the mixed number into an improper fraction below. I will cross cancel the 4's and sure enough, the answer is -11.

$$-\frac{11}{\cancel{4}_{1}} \times \frac{\cancel{4}^{1}}{1} = -\frac{11}{1}$$

Even if the x is on the right-hand side of the equation, it still works the same way. Look at this problem.

$$-84 = \frac{1}{2}x$$

23

To get x by itself, we need to get that $\frac{1}{2}$ off of it. Since they are multiplied, we can get rid of it by doing the opposite operation; division. I will divide the right-hand side by $\frac{1}{2}$, which will just equal x. And of course, whatever you do to one side you must do to the other side, so the EQUAtion stays EQUAL.

$$-84 \div \frac{1}{2} = x$$

Now all I have to do is solve the math. Since there is a negative number and a positive number, I know the answer will be negative. All that's left is turning 84 into a fraction, getting the reciprocal of $\frac{1}{2}$, and then multiplying straight across.

$$-\frac{84}{1} \times \frac{2}{1} = -\frac{168}{1} \qquad -168$$

But take another look at that last problem; $-84 \div \frac{1}{2}$. I'm not going to worry about the negative sign, for now, because I already know the answer will be negative. I am just looking at the math; 84 DIVIDED BY one half. So, let's think about that...if I had 84 sandwiches and then I DIVIDED them into half sandwiches, how many sandwiches would I have? The number of sandwiches would have doubled, right? Anytime a number is divided by one half, the number is doubled. That's why we ended up MULTIPLYING BY 2 to get the answer above. Get it?

To DIVIDE a number by $\frac{1}{2}$ is to DOUBLE it, but to MULTIPLY a number by $\frac{1}{2}$ is to cut the number in half.

Here is something to keep in mind about the things I'm teaching you in this book. Have you ever seen the movie "The Karate Kid?" In that movie, Daniel was taught several different skills; one at a time. He didn't understand what he was learning, but once his teacher put all his new skills together, he found out he could easily do several Karate moves. That's what we are doing here...well kind of. You are learning a bunch of different "algebra moves," so later you can solve big complicated math problems. Of course, they will only seem complicated to someone who hasn't learned all these rules, so make sure you learn the rule very well.

If you don't understand how to work with negative and positive numbers, then read my book called *Learn Math Fast System*. That book also covers how to work with mixed numbers. You must know how to work with mixed numbers before moving forward, otherwise you are just going to be even more confused. You don't want that.

If you are ready, complete the next worksheet.

Name: _____ Date: _____

WORKSHEET 3-2a

Solve for x.

1. $6x = 3$
2. $\frac{1}{2}x = 8$
3. $5x = 3.75$
4. $-5x = 25$

5. $-\frac{1}{8}x = 1$
6. $-2x = -34$
7. $-5 + x = 2\frac{1}{2}$
8. $10x = -6\frac{1}{4}$

9. $8x = 64$
10. $-7x = -49$
11. $16x = -4$
12. $72 = -9x$

13. $-3\frac{1}{8} + x = 2$
14. $8\frac{3}{8}x = -8\frac{3}{8}$
15. $12 + x = 4$
16. $x - \frac{5}{6} = \frac{2}{12}$

Try to solve these ones in your mind. (This is a tough one.)

17. $\frac{1}{2}x = -10$
18. $8 = \frac{1}{2}x$
19. $14x = 7$
20. $-2x = 4$

21. A 5-inch hamburger patty shrinks down to $\frac{3}{4}$ that size when cooked. Here is the math, $\frac{3}{4} \cdot 5 = x$ $\frac{15}{4} = x$. But Collin wants the cooked burgers to be exactly 3". What size hamburger patties should he make?

$$\frac{3}{4}x = 3 \text{ inches}$$

LESSON 3: SOLVE FOR X WITH MULTIPLICATION

In that last lesson, you learned how to "undo" multiplication by using division. This time you will learn how to "undo" division with multiplication. Sounds pretty easy, so let's get started.

$$\frac{x}{3} = 5$$

Can you read this one? That means, "x divided by 3 equals 5." To get x by itself, we need to get that 3 over to the other side. The x is *divided by 3*, so we need to do the opposite. The opposite is to *multiply by 3*.

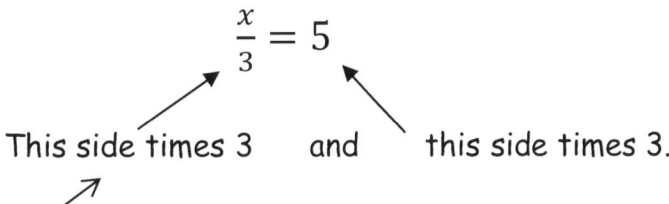

This side times 3   and   this side times 3.

When we multiply this side by 3, we are left with just an x...perfect. The only math left to do is on the other side, $5 \times 3 = 15$. Here is how you should write that problem out on paper (except for the arrows).

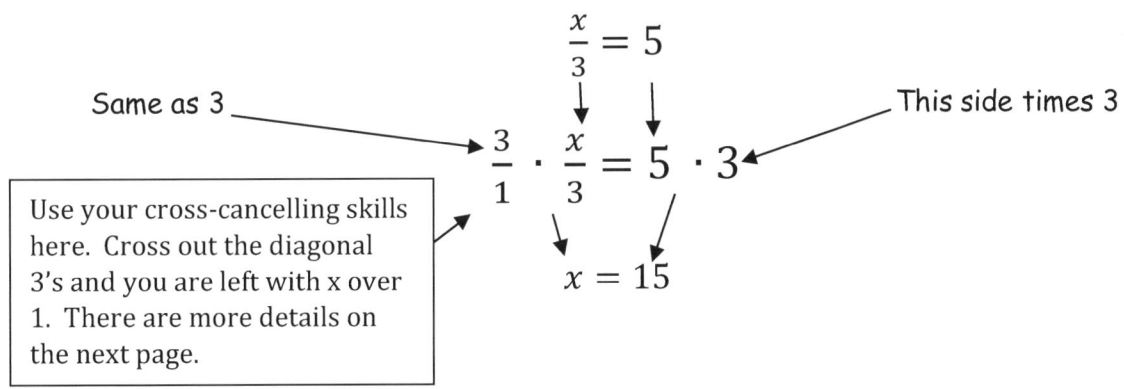

Same as 3 — This side times 3

Use your cross-cancelling skills here. Cross out the diagonal 3's and you are left with x over 1. There are more details on the next page.

Is 15 the right answer? Put it back into the equation in place of x.

$$\frac{15}{3} = 5$$

Yes, 15 is the right answer because $15 \div 3 = 5$.

I want to make sure you completely understand the math I did on the last page, so I'll rewrite it below. Focus your attention on the part of the problem that isn't in gray, while I explain it in detail below.

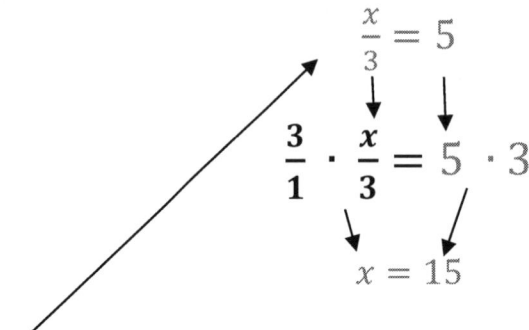

The original problem is, "x divided by 3 equals 5." The first step in algebra is to GET X BY ITSELF. Right now, "x" is not the only number on that side of the equal sign. It has "divided by 3" attached to it. How do you get rid of that 3? You must do the opposite operation - MULTIPLY by 3. To multiply $\frac{x}{3}$ by 3, we should write the number 3 as a fraction.

To write a whole number as a fraction, put it over the number 1. That is what I've done here.

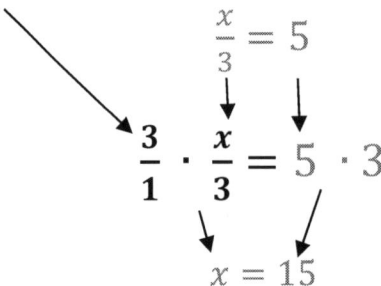

Since I am using the opposite operation, I know the answer will be x, but here is the math anyway. Use your cross canceling skills on this one.

$$\frac{\cancel{3}^1}{1} \cdot \frac{x}{\cancel{3}_1} = \frac{x}{1} = x$$

> X over 1 is the same thing as x because any number over 1 is that number.

Practice solving this type of math on the next worksheet.

Name: _____   Date: _____

WORKSHEET 3-3a

1. $\dfrac{x}{5} = 8$ 2. $\dfrac{x}{9} = 7$

3. $\dfrac{x}{10} = \dfrac{1}{2}$ 4. $\dfrac{x}{8} = 2\dfrac{3}{4}$

5. $\dfrac{x}{-8} = \dfrac{5}{8}$ 6. $\dfrac{7}{16} = \dfrac{x}{24}$

7. $-9 = \dfrac{x}{3}$ 8. $\dfrac{x}{4} = -3\dfrac{1}{2}$

9. $18 = \dfrac{x}{\frac{1}{2}}$ 10. $\dfrac{x}{3} = -4\dfrac{7}{8}$

11. Connor is playing in a championship football game. His team has a score of 56 points. Each touchdown was worth 7 points, so how many touchdowns did they make? Use the algebraic equation below to solve the problem.

$$7 = \dfrac{56}{x}$$

12. Keith is writing an essay for a contest. Each mistake is worth $-\frac{1}{2}$ points. If his total is $-5\frac{1}{2}$ points or more, he will win. How many mistakes can he make and still win? Use the algebraic equation below to solve the problem.

$$-\dfrac{1}{2} = \dfrac{-5\frac{1}{2}}{x}$$

LESSON 4: SOLVING FOR NEGATIVE X

Look at the next equation below.

$$100 - x = 10$$

What operation is being used? Subtraction, right? OK, then the opposite operation is addition. We need to add x to both sides of the equation or use the short cut by swinging the x over to the other side and changing the sign.

$$100 = 10 + x$$

OK, now what? We still don't have x by itself. We need to get rid of that 10 next. I will subtract 10 from both sides and the answer will be 90.

$$\begin{array}{r} 100 = 10 + x \\ -10 \;\; -10 \quad\;\; \\ \hline 90 \;\; = \;\; x \end{array}$$

But I want to show you another way to solve that problem. Since the 100 is a positive number and x is a negative number, we could have moved either one of those numbers over to the other side. This time I will subtract 100 from both sides.

$$\begin{array}{r} 100 - x = 10 \\ -100 \quad\;\; -100 \end{array}$$

100 – 100 is nothing, so it goes away. But this time, when you bring down the x, the negative sign has to stay with the x. I will do the math.

$$\begin{array}{r} 100 - x = \;\;\; 10 \\ -100 \quad\;\; -100 \\ \hline -x = -90 \end{array}$$

This is our answer, so far; negative x equals negative 90.

$$-x = -90$$

But we can't leave the answer like that. We aren't trying to solve for negative x, we want to solve for POSITIVE x.

$$-x = -90$$

No problem, that's easy to do because if negative x equals negative 90, then positive x equals positive 90. I can prove it, too.

$$-x = -90$$

Think of the equal sign as the center of the equation. Everything on the right side must equal everything on the left side. So, if we do something to one side, we must do it to the other side to keep it equal. We could add 10 to both side and they would still be equal or maybe add 1,000,000 to both sides. As long as we do the same thing to both sides, our equation will stay equal.

Watch what happens when we multiply both sides of this equation by -1.

$$-1 \cdot -x = -90 \cdot -1$$

A negative number, times a negative number will always have a positive answer. Look at the problem above. The signs are the same, so we know the answers will be positive.

$$-1 \cdot -x = -90 \cdot -1$$

$$-1 \cdot -x = x \qquad -90 \cdot -1 = 90$$

Any number times 1 is that number

$$x = 90$$

That proves if negative x equals negative 90, then positive x equals positive 90. If that was confusing, that's OK, just remember if -x = a negative number, then positive x = that number as a positive.

Here is another one. This one has a little more math to it, so let's use the short cut. The first step is to get x by itself. Start by moving the 44.

$$200 = 44 + 3x$$

Move the positive 44 over to the other side and change the sign.

$$200 - 44 = +3x$$

Do the math, $200 - 44 = 156$.

$$156 = 3x$$

We still need to get x by itself. Do the opposite, divide both sides by 3.

$$\frac{156}{3} = \frac{3x}{3}$$

Do the math, $156 \div 3 = 52$.

$$52 = x$$

Try this next one. It's a little more complicated, but you can do it.

$$-400 - 2x = -426$$

Do you know where to start? Do you swing over the -400? Or should you divide by -2? Well, if you were to divide by -2, you would have to divide the ENTIRE problem by -2 and that requires extra math that you will learn later on in this book.

Let's start with moving over the negative 400. The arrow is swinging the negative 400 to the other side with the opposite sign.

$$-400 - 2x = -426 + 400$$

$$-2x = -426 + 400$$

32

Do the math, $-426 + 400 = -26$.

$$-2x = -26$$

Do you know what to do next? To get x by itself, we need to get rid of the negative 2. Do the opposite operation by dividing both sides by -2.

$$\frac{-2x}{-2} = \frac{-26}{-2}$$
$$x = 13$$

There are two lessons to be learned here. First, if negative x equals a negative number, you can drop the two negative signs. And second, if there are two operations in one equation, then start with the addition/subtraction, not multiplication/division.

Try it yourself on the next worksheet.

Name: _____ Date: _____

WORKSHEET 3-4

Solve the following.

1. $93 - x = 47$
2. $x + 75 = -5$
3. $-x + 10 = 5$

4. $-5 + x = 11$
5. $x - (-23) = -43$
6. $-17 + x = -34$

7. $-3x = 9$
8. $12x = -48$
9. $99 = 11x$

10. $5x = 105$
11. $16x = 176$
12. $9x + 5 = 50$

13. $11x - 6 = 115$
14. $8 + 15a = 38$
15. $-33 - 14x = -145$

16. $45x - -2 = 92$
17. $\frac{1}{5}x + 7 = 11$
18. $x - \frac{5}{8} = 2\frac{3}{8}$

LESSON 5: SOLVING FOR X WITH DIVISION

Keep these two facts in mind as you read this next lesson.

$$121 \div 11 = 11$$
$$11 \div 1 = 11$$

I will rewrite them in a different format. The problems are the same, but I've replaced one number in each equation with an x.

$$\frac{x}{11} = 11 \qquad \frac{11}{x} = 11$$

The first problem is saying, $x \div 11 = 11$. The second problem is saying, $11 \div x = 11$. Do you see the difference? One is divided by 11, the other one is divided by x. In order to get x by itself, we have to do the opposite math. Let's look at just the first problem.

$$\frac{x}{11} = 11$$

This one is "divided by 11," so we need to do the opposite; multiply both sides by 11. When I multiply the left-hand side of the equation by 11, it turns into just x because we "undid" it. When I multiply the other side of the equation by 11, I get 121. I have written all my work below. Notice that the original problem is written in gray.

$$\frac{\cancel{11}^{1}}{1} \cdot \frac{x}{\cancel{11}_{1}} = 11 \cdot 11$$

$$\frac{x}{1} = 121$$

$$x = 121$$

After canceling the 11's we are left with $\frac{x}{1}$ which is the same thing as x.

35

In the second problem, we have to multiply both sides by x. I've done the math below. Read through it to make sure I got it right.

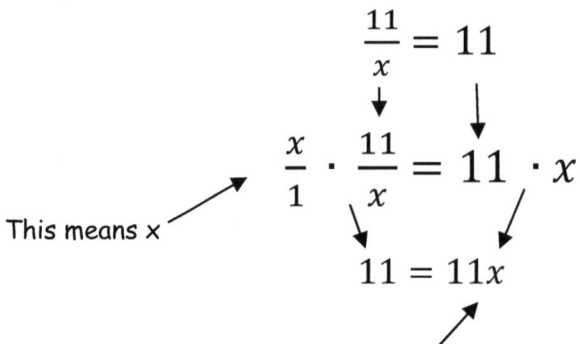

This means x

$$\frac{11}{x} = 11$$

$$\frac{x}{1} \cdot \frac{11}{x} = 11 \cdot x$$

$$11 = 11x$$

We still don't have x by itself. This x has an 11 attached to it with multiplication. Do the opposite operation to get x by itself.

Divide by 11 $\longrightarrow \quad \dfrac{11}{11} = \dfrac{11x}{11} \quad \longleftarrow$ Divide by 11

$$1 = x$$

Let's try that problem again using the short cut. This problem is "11 divided by x." The opposite is to multiply by x. Just swing the x over to the other side and change it to multiplication; doing that will turn this side into 11.

$$\frac{11}{x} = 11 \cdot x$$

We can get rid of the dot because 11x is the same thing as 11 · x. Our new problem looks like this:

$$11 = 11x$$

Divide both sides by 11, to get x by itself.

$$\frac{11}{11} = \frac{11x}{11}$$

$11 \div 11 = 1 \qquad\qquad 11x \div 11 = x$

$$1 = x$$

36

We just solved that problem two ways: the long way and the short cut way. Next, we will solve it the logical way. Look at the problem one more time.

$$\frac{11}{x} = 11$$

Can you write the number 11 as a fraction? Any whole number can be written as a fraction by putting the number 1 under it. So, the number 11, written as a fraction, looks like this:

$$\frac{11}{1} = 11$$

Look at the two problems above. Do you see why x = 1? Solving this problem logically is definitely the fastest way, but a lot of math teachers like to see the steps written out, so you need to know the long way, too. Look at this next problem.

$$\frac{15}{x} = 3$$

Can you solve this one on your own? First read the equation. It says, "Fifteen divided by x equals three." Think about that logically. Fifteen divided by *what* will equal three? Fifteen divided by five equals three, so I know the answer is 5. But just to make math teachers around the world happy, let's show our work.

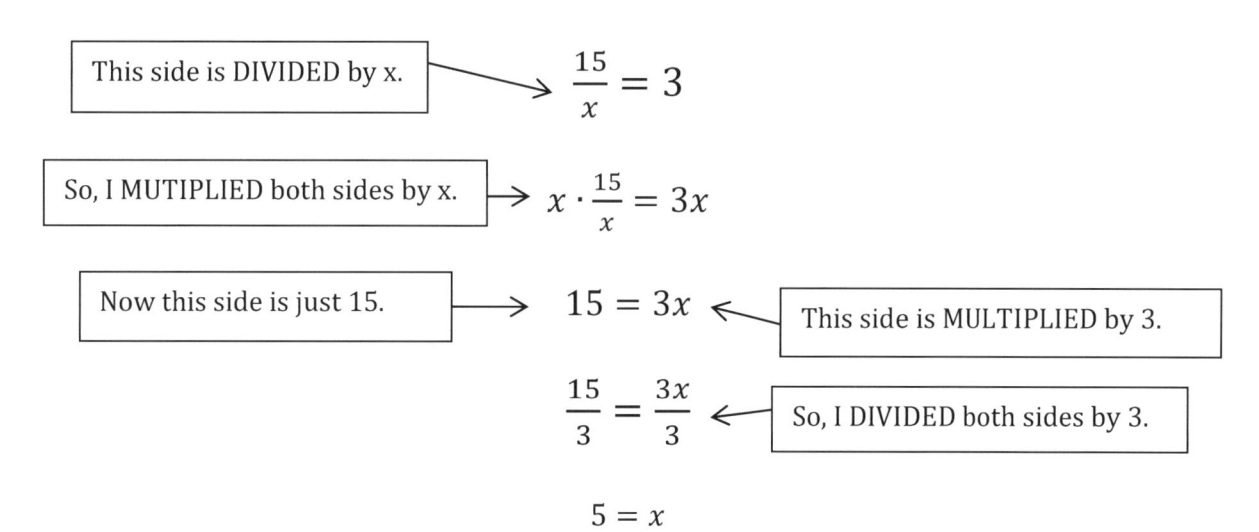

| This side is DIVIDED by x. | → | $\frac{15}{x} = 3$ |

| So, I MUTIPLIED both sides by x. | → | $x \cdot \frac{15}{x} = 3x$ |

| Now this side is just 15. | → | $15 = 3x$ | ← | This side is MULTIPLIED by 3. |

$$\frac{15}{3} = \frac{3x}{3}$$ ← So, I DIVIDED both sides by 3.

$$5 = x$$

Complete the next worksheet. Try to solve each one the easiest way you can.

Name: _____   Date: _____

WORKSHEET 3-5

Solve the following.

1. $\frac{54}{x} = 9$ 2. $\frac{x}{8} = 7$ 3. $\frac{48}{x} = 8$

4. $\frac{x}{5} = 4$ 5. $\frac{144}{x} = 12$ 6. $\frac{56}{x} = 28$

7. $\frac{28}{x} + 3 = 10$ 8. $8 + \frac{56}{x} = 15$ 9. $\frac{x}{6} - 2 = 4$

10. $\frac{1}{3}x = 2$ 11. $\frac{200}{x} = 1$ 12. $\frac{1}{4}x = 4$

13. $\frac{x}{2} - 2 = 3$ 14. $\frac{96}{x} + 9 = 17$ 15. $\frac{14}{x} = 14$

LESSON 6: WHAT IS Y?

Now that you have a basic understanding of how to use an x in a math equation, let's put a "y" into the equation.

I told you that "x" is a question mark because "x" is an unknown number. Math people call "x" a *variable* because it can *vary*. One time it's a 3, another time it's an 11; it *varies*. Look at this problem.

$$x + x = 6$$

In this problem, x has to be 3 because x can only be one number in each equation. We wouldn't say x = 4 and x = 2 in the same problem because x can only equal one thing. If we want two different unknown numbers, we would have to use another letter. We will use y. It doesn't matter which letter of the alphabet I use; they are all called variables in algebra.

$$x + y = 6$$

In this problem, if we want to solve for x, we need get rid of the y. To move the y over to the other side, we have to subtract it. Or just swing it over and change the sign.

$$x + y = 6$$
$$x = 6 - y$$

That is the answer. That's all we can do, since we don't know what y is. That math is called "solving for x in terms of y." It means, solve for x like usual, but the answer is going to have the letter y in it somewhere. Let's try that same problem only this time, we will "solve for y in terms of x."

$$x + y = 6$$
$$y = 6 - x$$

What we just did is called "solving for y in terms of x." That means, I can tell you how much y is as soon as you tell me how much x is.

Let's try one more. This time we will solve for y in terms of x.

$$\frac{x}{y} = 5$$

Start by "undoing" the division. Multiply each side by y. Here is what you'll get:

$$x = 5y$$

That is solved for x, but we want to solve for y. We need to move that 5 over to the other side. Do the opposite operation. Divide each side by 5.

$$\frac{x}{5} = \frac{5y}{5}$$

Now we have an answer for y in terms of x.

$$\frac{x}{5} = y$$

But, of course, I have an easier way. Let me explain. Do you believe these two equations? Of course, you do.

$$15 \div 3 = 5$$
$$15 \div 5 = 3$$

Now I am going to rewrite those exact same equations a little differently.

$$\frac{15}{5} = 3 \qquad\qquad \frac{15}{3} = 5$$

Do you see how the 5 and the 3 can be switched around and the equation is still true? So, when you see a problem like the one below, certainly you could just switch the "x" and the "5" and the problem would be solved for x.

$$\frac{y}{x} = 5 \qquad\qquad \frac{y}{5} = x$$

40

Now that's the fastest way yet!

Try some on your own. If you have problems, look at the answers to help you understand. Remember, when you are solving for x in terms of y, your answers will be x = something with a y. When you are solving for y in terms of x, your answer will be y = something with an x.

Name: _____   Date: _____

WORKSHEET 3-6

Solve for x in terms of y.

1. $x + y = 24$
2. $y + x = 18$
3. $x + 8 = y$

4. $x - 4 = y$
5. $xy = 30$
6. $\dfrac{x}{y} = 15$

7. $17 + x = y$
8. $\dfrac{y}{x} = 16$
9. $\dfrac{x}{y} = 77$

Solve for y in terms of x.

10. $\dfrac{48}{y} = x$
11. $\dfrac{x}{y} = 4$
12. $\dfrac{144}{y} = x$

13. $5y = x$
14. $9x + y = 50$
15. $11x + y = 115$

16. $x = 100y$
17. $x = 41y$
18. $x = y - 10$

LESSON 7A: RATIOS

Now that you have a handful of algebra skills, let me show you how you can use them. But first, you must learn about *ratios*. Do you remember learning that a fraction is actually a division problem? For example, look at the fraction below.

$$\frac{1}{4}$$

This fraction is read as "one fourth." But it can also be read as, "one divided by four." Go ahead and do the math, it will equal .25 because $\frac{1}{4}$ is the same thing as .25.

In this lesson, you will learn one more way to read that same fraction. Not only is it a division problem, but it is also called a *ratio (ray-she-oh)*. To read that fraction as a ratio, you would say, "one to four." It can also be written as 1:4 and it, too, is read as "one to four."

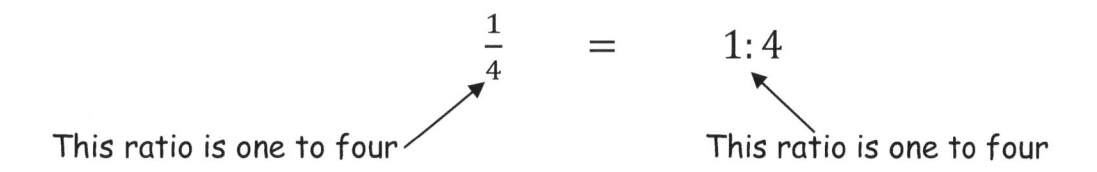

$$\frac{1}{4} \quad = \quad 1:4$$

This ratio is one to four   This ratio is one to four

OK, so you can read a ratio, right? Now I'll tell you what it means. Let's say you are making a ball of dough. The directions say to use 3 cups of flour and 2 cups of water. That ratio of flour to water would be written as:

$$3:2 \quad \text{or} \quad \frac{3}{2}$$

Well, now let's say you want to make a HUGE ball of dough. With this ratio you could use any unit of measurement. You don't have to use a measuring cup to make this dough. You could use 3 BUCKETS of flour and 2 BUCKETS of water. Or maybe you want to make a little bit of dough. You could use 3 teaspoons of flour to 2 teaspoons of water.

As long as the unit of measurement is the same for both numbers, you can drop the unit and change it to whatever you need it to be. Writing numbers in a ratio leaves us flexible to change the unit of measurement and still be able to make perfect dough.

We use ratios in math to compare two numbers. Below are a few examples.

Have you ever been to a big wedding? Did you see a set of people standing up next to the bride and groom? At a typical wedding you might see three girls standing next to the bride and three men standing next to the groom. The number of girls and boys are usually the same, so they can walk down the aisle together in equal pairs. The ratio of men to women in this wedding party would be 3:3. I bet the bride would not want the ratio to be 1:3. That would look funny, don't you think? So, let's write that ratio, 3:3 as a fraction.

$$\frac{3}{3}$$

Can you reduce that fraction? Of course, you can! It reduces down to $\frac{1}{1}$. That means there is to be 1 boy for every 1 girl. Ratios are reduced for this reason, because you don't HAVE TO have 3 boys and 3 girls in each wedding. You just need to stick to the ratio of 1:1.

Now let's say Grandma has a big bowl of candy. She wants to pass out one piece of candy to each kid. That's a ratio of 1:1. The kids were hoping she would pass out the candy with a ratio of 3:1. That's three pieces of candy per kid. How much candy would each kid get if she passed out the candy with the ratio below?

$$\frac{1}{3}$$

That's right, each kid would have to share a piece of candy giving each person a 1/3 piece.

Here is another example. At a daycare, the rule is to have one adult for every seven kids. They wouldn't expect one adult to take care of 21 kids and they wouldn't want to pay five adults to take care of one child, so they keep the ratio at 1:7. That gives each child enough attention without the adult feeling overloaded.

I have another situation that calls for ratios. A boat has an outboard motor on the back to push it around in the water. The motor uses a mixture of gas and oil to run properly. Often times the mixture is said to be "2 parts gas to 1 part oil." How would you write that as a ratio?

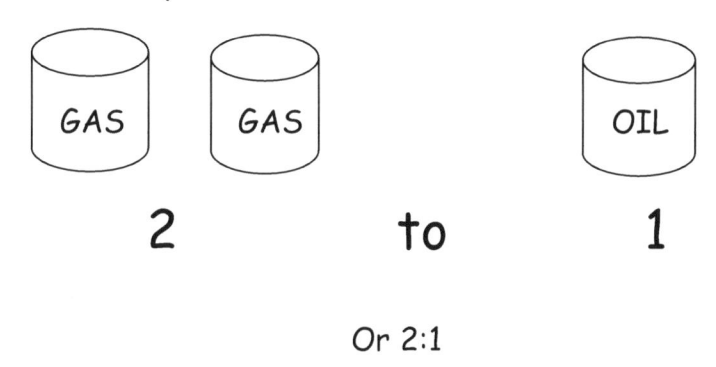

| GAS   GAS   OIL |

2   to   1

Or 2:1

You could use 2 GALLONS of gas to 1 GALLON of oil. You could also use 2 quarts of gas to 1 quart of oil. As long as you keep the ratio at 2:1, the motor will run smoothly.

I'll give you another one. When I drink a glass of water, I like it to be cold. If I put one ice cube in a 12-ounce glass of water, the ice will probably melt before the water gets as cold as I'd like it to be. If I add 20 ice cubes, not only will I need a bigger glass, but it will be difficult to drink because the RATIO of ice cubes to ounces of water will be off. How would you write a ratio to describe the amount of ice to water in the first glass with only one ice cube?

$$\frac{1 \; ice \; cube}{12 \; ounces \; of \; water}$$

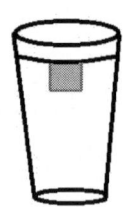

The ratio is 1:12. That's 1 ice cube per 12 ounces of water.

How about the second glass that had 20 ice cubes in it? Can you write that ratio?

$$\frac{20 \; ice \; cube}{12 \; ounces \; of \; water}$$

That ratio is 20:12. Let's reduce that down by dividing both numbers by 4.

$$\frac{5}{3}$$

This ratio tells us that there are 5 ice cubes for every 3 ounces of water. That's more ice than water! That's too much ice for my taste.

I think a more appropriate ratio would be about 4:12. I'll write that as a fraction and then reduce it.

$$\frac{4}{12} = \frac{1}{3}$$

There we go. That's much better; one ice cube per 3 ounces of water. Now that's refreshing!

In a typical public-school classroom, there is one teacher and about 30 kids. In that case, the ratio of teachers to students would be 1:30. If I asked you the ratio of KIDS to TEACHERS, you would have to say 30:1. Do you see how the words have to line up with the numbers? What is the ratio of teachers to students in your math class? I homeschooled my only child, so our ratio was 1:1.

Complete the next worksheet to get a little practice writing a ratio and then we'll get to working with them.

Name: _____   Date: _____

WORKSHEET 3-7a

Write each ratio as a fraction.

1. The ratio of 3 to 7.

2. The ratio of 5 to 10.

3. The ratio of 8 to 3.

4. The ratio of $\frac{1}{2}$ to 3. (Don't be afraid to have a fraction as a numerator)

5. A fisherman needs 12 pounds of weight for every 60 feet of fishing line. What is the ratio of weight to feet? Reduce your answer.

6. At the school dance, there were 40 boys and 20 girls. What is the ratio of boys to girls? Reduce your answer.

7. The votes were counted. There were 6 "yes" votes and 18 "no" votes. What is the ratio of yes to no votes? Reduce your answer.

8. The campers received one tent per four campers. What is the ratio of campers to tents? Write your answer in this format: 9:3.

9. A race car travels three miles in one minute. Another car drove one mile in one minute. Write a ratio that compares the distance of the race car to the other car.

10. A recipe for potato salad suggests using 3 potatoes for each serving. What is the ratio of potatoes per person?

LESSON 7B: WORKING WITH RATIOS

Now that you have had some practice writing ratios, I'll show you how to use them. You should try to remember this process for the rest of your life because you will probably have to use it at some point, even if you don't grow up to be a mathematician.

In the last lesson, we used ratios to compare things such as the number of boys to girls, or the amount of flour to water. That's one way to use ratios. Now I'll show you another one.

I mentioned that we can drop the unit of measurement, if they are the same for both numbers in a ratio. But often times, the units *are* different. No problem, just read the line in the fraction as the word "per." For example, here is a ratio of cost per books. It costs $63.00 for 7 books.

Read this line as "per" $\longrightarrow \dfrac{\$63}{7\ books} = \$9$ per book

If you do the math, $63 \div 7$, you will get the price per book. Notice that our math was "dollars divided by books." Your answer will be said the same, just replace "divided by" with "per." The answer will be "dollars per book."

Here is another way to use a ratio. Let's say you are making hamburgers at a bar-b-que party. At the end of the day, you made a total of 36 hamburgers. There were 20 people at the party. The ratio of hamburgers per person is written below.

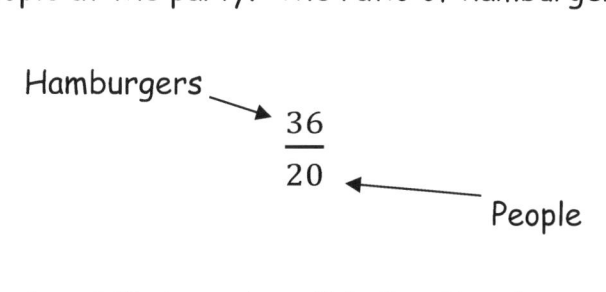

Hamburgers $\longrightarrow \dfrac{36}{20} \longleftarrow$ People

Now use your reducing skills to reduce this fraction down to the smallest possible denominator. Divide each number by 4; that will reduce it.

49

$$\frac{36}{20} = \frac{9}{5}$$

Now that the ratio is reduced, let's do the division. It is "burgers divided by people," so the answer will be "burgers per person."

```
         1.80 hamburgers per person
      _____
   5) 9.00
      5
      ‾‾
      40
```

The next time you have a bar-b-que, you will know how many hamburgers to make. Just multiply the number of people by 1.8 because our ratio from the previous party tells us that we should make 1.8 hamburgers per person.

Another way to use ratios is to find the price per unit, let me explain. Let's say you have the option of buying a small pack of gum or a large pack of gum. You want to get the best price, so let's find out how much you are paying for each piece of gum. That way you will know which one the better deal is.

The big pack costs $2.99 for 50 pieces. The small pack costs $0.99 for 10 pieces of gum. I will write them both as a ratio. We need to keep the units the same for each ratio, so I will write them both in cents not dollars.

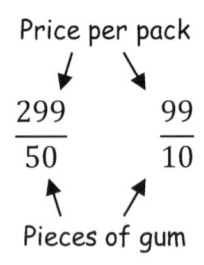

Price per pack
↓ ↘
$$\frac{299}{50} \quad \frac{99}{10}$$
↑ ↗
Pieces of gum

Do the division of "cents divided by pieces" and your answers will be the price per piece of gum, in cents.

The big pack of gum	The small pack of gum

$$\begin{array}{r} 5.98 \text{ cents per piece} \\ 50 \overline{)299.00} \\ \underline{250} \\ 490 \\ \underline{450} \\ 400 \end{array} \qquad \begin{array}{r} 9.9 \text{ cents per piece} \\ 10 \overline{)99.0} \\ \underline{90} \\ 90 \\ \underline{90} \end{array}$$

With ratios, we can see that the big pack of gum is the better value. Let me give you a few more examples of ratios and how to write them.

An ice cream cone is built with 2 scoops of ice cream per cone. Look at the line as the word "per."

$$\frac{2 \text{ scoops}}{1 \text{ cone}} = 2 \text{ scoops per cone}$$

Do the division, 2 ÷ 1, to find how many scoops there are per cone. OK, that one was obvious, so let's try a more challenging ratio.

The car drove 98 miles in 2 hours. How many miles did it drive per hour? Write it as a ratio, 98 miles per 2 hours, and then divide.

$$\frac{98 \text{ miles}}{2 \text{ hours}} = 49 \text{ miles per hour}$$

A college fund has $75,000. This fund needs to pay for 3 girls to go to college. How much will each girl get? Write it as a ratio and then divide.

$$\frac{\$75{,}000}{3 \text{ girls}} = \$25{,}000 \text{ per girl}$$

I put gas in the car. It cost $39.80 for 11 gallons of gas. How much did I pay per gallon? Write a ratio and then divide. I show my work on the next page.

$$\frac{\$39.80}{11 \text{ gallons}} = \$3.61 \text{ per gallon}$$

```
         3.61 9/11
      _____
   11) 39.80
       33
       ──
        68
        66
        ──
        20
        11
        ──
```

You've probably noticed by now that all we are doing is dividing. By writing a ratio first, you can tell which number to divide into which.

Try some on your own on the next worksheet.

Name: _____ Date: _____

WORKSHEET 3-7b

1. Write a ratio that shows 14 dog bones for 7 dogs. Then show how many dog bones per dog.

2. Write a ratio to help you calculate the cost per can of pop. A 6-pack of pop costs $2.94. How much per can? Hint: your answer will be in cents, so your ratio should be in cents not dollars.

3. Write a ratio, using the ratio symbol ":" to show 2 teachers to 44 students.

4. Write two ratios to help you figure out which is the better value. You can spend $249 for 100 t-shirts or you spend $49 to buy 10 t-shirts. Find the price per t-shirt. Which is the better deal?

5. We traveled 444 miles in 3 days. How many miles did we travel per day?

6. Write three different phrases to describe this: $\frac{1}{3}$

   _____   _____   _____

7. It cost $46.05 for 15 gallons of gas. Write a ratio and then solve it to find the price per gallon.

8. Pat ran 3 miles in 20 minutes. How many miles did she run per minute?

9. Flo spent $11.60 on 5 pounds of hamburger. How much did she pay per pound of hamburger?

LESSON 8A: PROPORTIONS

Now we will use your ratio knowledge to do some cool algebra tricks. Below is a ratio, and I have reduced it down.

$$\frac{3}{18} = \frac{1}{6}$$

Look again at those two ratios above. Whenever you have a "this ratio equals that ratio" kind of equation, you have a PROPORTION. Proportions are fun because they are easy to solve and can be very useful, too.

The first thing you should notice about a proportion equation is that when you cross multiply the ratios in the direction of the arrows below, you will get the same answer in both directions because they are equivalent (equal) ratios (or fractions).

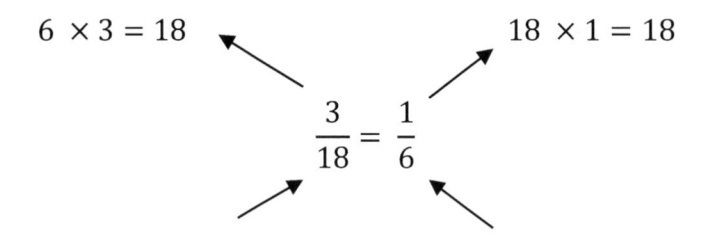

$$6 \times 3 = 18 \qquad 18 \times 1 = 18$$

$$\frac{3}{18} = \frac{1}{6}$$

It works like that with any two equivalent fractions or ratios. If you cross multiply in the direction of the arrows above, you will get the same answer. If you don't get the same answer, then the fractions are not equal. And on a side note, if the fractions aren't equal, the fraction with the larger answer above it is the bigger fraction.

With that knowledge, mathematicians figured out a cool trick. Look at the two ratios below. One of them has a missing a number. I put an "x" in place of the missing number.

$$\frac{12}{x} = \frac{6}{5}$$

To solve this proportion equation, cross multiply and then use algebra to solve for x.

$$12 \cdot 5 = 6x$$

$$60 = 6x$$

$$\frac{60}{6} = \frac{6x}{6}$$

$$10 = x$$

This time try it on your own. I'll give you two equivalent ratios. Cross multiply and then use algebra to solve for x.

$$\frac{20}{5} = \frac{8}{x}$$

Start by cross multiplying: $40 = 20x$

Divide each side by 20 to get x by itself: $\frac{40}{20} = \frac{20x}{20}$

Do the math to solve for x: $2 = x$

Let's see if I got the right answer. I'll replace the x in the proportion above with my answer. Look at the two ratios below. The first one is 20 ÷ 5 and the second one is 8 ÷ 2. They both equal 4, so my answer is correct.

$$\frac{20}{5} = \frac{8}{2}$$

Stop here, and solve this one on your own. Cross multiply to solve for x.

$$\frac{x}{30} = \frac{3}{5}$$

Did you get x = 18? If you did, complete the next worksheet. If you didn't, go back a few lessons and try again.

Name: _____   Date: _____

WORKSHEET 3-8A

Cross multiply and then solve for x.

1. $\dfrac{40}{x} = \dfrac{5}{1}$

2. $\dfrac{x}{16} = \dfrac{2}{8}$

3. $\dfrac{54}{x} = \dfrac{6}{1}$

4. $\dfrac{6}{36} = \dfrac{3}{x}$

5. $\dfrac{24}{12} = \dfrac{x}{2}$

6. $\dfrac{105}{5} = \dfrac{42}{x}$

7. $\dfrac{\frac{2}{5}}{x} = \dfrac{10}{50}$

Solve for x.

8. 7 is to 21 as x is to 6. $\left(\dfrac{7}{21} = \dfrac{x}{6}\right)$

9. 15 is to x as 30 is to 6.

10. x is to 100 as 9 is to 30

11. 7 is to 2 as 14 is to x.

LESSON 8B: UNIT CONVERSION

Now for the fun stuff: Unit Conversion. I know that sounds awful, but it's not. The word "conversion" means to convert something. Have you ever heard of a convertible car? A convertible car "converts" from a normal car to a car without a roof.

Have you seen the movie *Transformers*? In that movie, robots convert into vehicles, or maybe the vehicles are converting into robots. I'm not sure, but you get the idea, right? To convert something is to turn it into something else. This lesson will be about converting *units*. When I say units, I mean feet, miles, cups, hours, centimeters, dollars, pounds, or any other unit of measurement. For example, we might convert two meters into feet, or three miles into kilometers. That's what *Unit Conversion* means, and it's all done with proportions. Let me explain.

Do you know how many minutes are in an hour? There are 60, right? I will write down the ratio of one hour to 60 minutes.

$$\frac{1 \text{ hour}}{60 \text{ minutes}}$$

Now let's say I want to know how many minutes are in 8 hours. I can write a PROPORTION to figure that out. The equation below is read as "One hour is to 60 minutes as 8 hours are to x minutes."

$$\frac{1 \text{ hour}}{60 \text{ minutes}} = \frac{8 \text{ hours}}{x \text{ minutes}}$$

Just make sure your hours and minutes are located in the same place in each ratio. All we do now is cross multiply to get our answer. So, let's see that's 1 times x, equals 8 times 60.

$$1x = 480$$

There are 480 minutes in 8 hours. Let's try another easy one like that.

Do you know how many feet are in a mile? There are 5,280 feet in a mile. So, how many MILES are in 13,200 feet? I will write a proportion to figure that out.

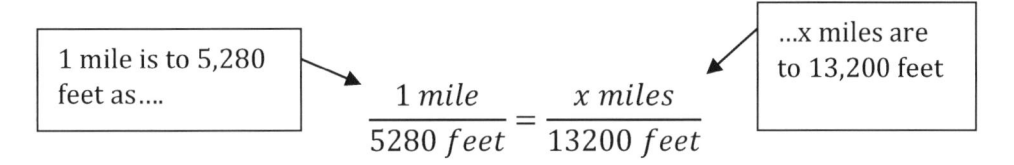

> 1 mile is to 5,280 feet as....

> ...x miles are to 13,200 feet

$$\frac{1\ mile}{5280\ feet} = \frac{x\ miles}{13200\ feet}$$

Notice that I put the "x" up there in the "miles" location. Now all we have to do is cross multiply to find out how many miles there are in 13,200 feet.

$$13,200 = 5,280x$$

$$\frac{13200}{5280} = x$$

> Divide both sides by 5280.

$$2.5 = x$$

That is our answer. There are 2 ½ miles in 13,200 feet.

I would like to have said that I put the "x" in the numerator, but that word is saved for fractions. Even though these ratios LOOK like fractions and ACT like fractions, in this situation they aren't. They are two ratios, which equal each other, and that creates a type of equation called a proportion.

So, if we can't use the words numerator and denominator, what do we call each location of these ratios? I'm glad you asked. They have their own special names. Look at the proportion below.

$$\frac{5}{8} = \frac{10}{16}$$

These two numbers are called the *extremes*. The 8 and 10 are called the *means*. Mathematicians like to say, "The product of the means equals the product of the extremes." That is true, but it doesn't help me remember which one is which. I know the diagonals both have the same name, but it's sometimes difficult to remember which set of diagonals are the means and which are the extremes, so I made this little trick.

$$\frac{Extreme}{Means} = \frac{Means}{Extreme}$$

It is "Extremely Mean" to make us remember these terms. Whether you read from top to bottom or from left to right, it still says, "Extreme(ly) mean." That should help you remember which one is which.

Now let's get back to the Conversion of Units. Those last examples were kind of easy. You could have probably figured out the answers without using proportions, so let's make this a little more difficult.

Do you know how many pounds are in one kilogram? I don't either, but I have a chart that tells me there are approximately 2.2 pounds in one kilogram. I'll write that in a ratio.

$$\frac{2.2 \; pounds}{1 \; kilogram}$$

Now let's figure out how many kilograms there are in 8 pounds. Remember to keep the "pounds" and "kilograms" in the same place. Before you look at the proportion below, try to guess where the "x" should go. Remember: 2.2 pounds are to 1 kilogram as 8 pounds are to how many kilograms.

$$\frac{2.2 \; pounds}{1 \; kilogram} = \frac{8 \; pounds}{}$$

That's right, we are looking for x kilograms, so the "x" is the extreme.

$$\frac{2.2 \; pounds}{1 \; kilogram} = \frac{8 \; pounds}{x \; kilograms}$$

All that's left is the fun part, cross multiplying.

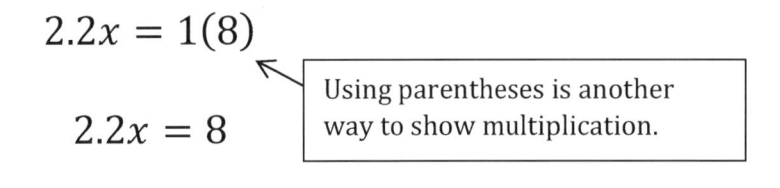

$$2.2x = 1(8)$$

Using parentheses is another way to show multiplication.

$$2.2x = 8$$

$$x = \frac{8}{2.2}$$

$$x = 3.64$$

The answer is 3.64 kilograms. Eight pounds is equal to 3.64 kilograms. (You should know that this answer is rounded. The conversion between kilograms and pounds is not exact.)

Let's try another one. My friend, Teresa, said she is going to run a 50 Kilometer race. In America we use "miles" more often than "kilometers," so I'm not sure how far she is going to run. Let's figure it out. According to my sources, there are .62 miles in one Kilometer, so how many miles are in 50 Kilometers?

$$\frac{1\ K}{.62\ m} = \frac{50\ K}{x\ miles}$$

Try to solve this one on your own. Go ahead, I'll wait. This is the answer I got.

$$x = 31$$

Teresa will be running 31 miles during her 50K race.

I was told to bring 5 quarts of coffee to the potluck. My coffee pot only has markings for liters. How many liters of coffee should I bring? I know there are 1.06 quarts in 1 liter, so let's start there.

| 1 liter is 1.06 quarts. | → | $\dfrac{1}{1.06} = \dfrac{x}{5}$ | ← | How many liters in 5 quarts? |

$$5 = 1.06x$$

$$4.72 = x$$

I should make 4.72 liters of coffee. Try some on your own.

Name: _____   Date: _____

WORKSHEET 3-8B

Use Proportions to solve the following.

1. There are 3 feet in 1 yard. How many yards are in 120 feet?

2. There are 12 inches in 1 foot. How many inches are in 8 feet?

3. There are 16 ounces in 1 pound. How many pounds in 128 ounces?

4. There are 3.28 feet in 1 meter. How many feet in 3 meters?

5. One gram equals .035 ounces. 14 grams equal how many ounces?

6. There are 2.54 centimeters in 1 inch. 5 inches equal how many centimeters?

7. There are 8 ounces in 1 cup and 16 cups in 1 gallon. How many ounces are in 1 gallon?

LESSON 9: EXPONENTS^2

Below is an example of a number with an exponent. In this example, the number 4 is called the *base*, and the little 2 is the *exponent*.

$$4^2$$

When you put an exponent of 2 above a number, it is called squaring a number. To square a number is to multiply it by itself. For example, 4^2 means 4 × 4 or 16.

3^2 *means* 3 × 3 *or* 9
2^2 *means* 2 × 2 *or* 4

Can you solve this problem? It is read, "two squared plus three squared."

$$2^2 + 3^2 =$$

Solve each one separately and then add them together.

$$2^2 + 3^2 =$$
$$4 + 9 = 13$$

Remember not to multiply the base number by 2. It is the base number times itself.

Complete the next worksheet. It's easy.

Name: _____   Date: _____

WORKSHEET 3-9

1. $6^2 =$ 2. $7^2 =$ 3. $8^2 =$

4. $3^2 =$ 5. $2^2 =$ 6. $9^2 =$

7. $10^2 =$ 8. $5^2 =$ 9. $4^2 =$

10. $11^2 =$ 11. $1^2 =$ 12. $12^2 =$

13. $3^2 + 3^2 =$ 14. $2^2 + 4^2 =$ 15. $5^2 + 3^2 =$

16. $6^2 + 4^2 =$ 17. $8^2 + 7^2 =$ 18. $3^2 + 2^2 =$

19. $4^2 \cdot 2 =$ 20. $5^2 \cdot 4 =$ 21. $6^2 \cdot x =$

22. $6^2 \cdot \frac{1}{2} =$ 23. $4^2 \div \frac{1}{2} =$ 24. $2^2 \cdot 3^2 =$

LESSON 10: MORE EXPONENTS[3]

Look at this exponential expression. In this example, the *base* is the number 4. The little 3 is the *exponent*.

$$4^3$$

When the exponent is a 3, it is called *cubed*, instead of squared. When the exponent is a 3, the number shows up 3 times in a multiplication problem.

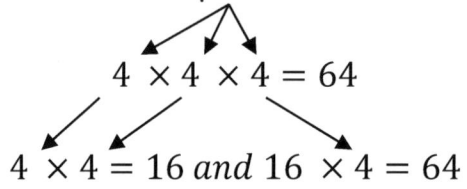

$$4 \times 4 \times 4 = 64$$

$$4 \times 4 = 16 \text{ and } 16 \times 4 = 64$$

Here is another one. Can you solve this exponential expression?

$$5^3 =$$

This problem means 5 x 5 x 5. The answer is 125 because 5 x 5 = 25, and 25 x 5 = 125.

Can you solve this problem?

$$2^3 + 3^2 =$$

Be careful. Make sure you read the exponents correctly. Here is the math.

$$2^3 + 3^2 =$$

$$2 \times 2 \times 2 = 8 \qquad 3 \times 3 = 9$$

$$8 + 9 = 17$$

Be careful not to make the same mistake everyone makes when first learning exponents. Make sure you are multiplying the big number by itself. Don't multiply the big number by the little number. That's a common mistake.

When you read the number 4^2, it is read, "Four squared." Anytime there is an exponent of 2 above a number, it is read as squared. When you multiply a number by itself, it is called squaring the number.

When you read the number 4^3, it is read, "Four cubed." Anytime there is an exponent of 3 above a number, it is read as cubed.

The number 4^4 is read, "Four to the fourth power." 4^{10} is read, "Four to the tenth power." It works like that for any exponent besides 2 and 3; they are always read as squared and cubed. You will learn why in Volume IV.

Before I have you complete the next worksheet, I want to teach you something else. Have you ever heard someone say that something "grows exponentially?" It's a mathematical phrase. Let me explain. Watch how the answers to the exponential problems below get larger and larger.

$$5^2 = 25$$
$$5^3 = 125$$
$$5^4 = 625$$
$$5^5 = 3,125$$
$$5^6 = 15,625$$
$$5^7 = 78,125$$
$$5^8 = 390,625$$
$$5^9 = 1,953,125$$
$$5^{10} = 9,765,625$$

These numbers are more than doubling each time, they are getting exponentially bigger.

When a number is growing rapidly, like the numbers above, we say they are getting exponentially bigger. They are not just doubling or tripling each time. They are getting *exponentially* larger.

OK, now you can complete the next worksheet.

Name: _____   Date: _____

WORKSHEET 3-10

1. $6^3 =$ 2. $7^3 =$ 3. $8^3 =$

4. $3^3 =$ 5. $2^3 =$ 6. $9^3 =$

7. $10^3 =$ 8. $5^3 =$ 9. $4^3 =$

10. $11^3 =$ 11. $1^3 =$ 12. $12^3 =$

Solve the following. Read the exponents carefully.

13. $3^3 + 3^2 =$ 14. $2^3 + 4^3 =$ 15. $5^3 + 3^2 =$

16. $6^2 + 4^3 =$ 17. $8^3 + 10 =$ 18. $3^3 + 2^3 =$

19. $5^3 \cdot 2 =$ 20. $5^3 \cdot 3 =$ 21. $4^3 \cdot y =$

22. $2^3 \cdot \frac{1}{4} =$ 23. $3^3 \div \frac{2}{7} =$ 24. $4^3 \cdot 7^2 =$

LESSON 11: SQUARE ROOT

By now you should know that the opposite of addition is subtraction. By that I mean if you add 2 rocks to a bucket of rocks, you can undo that by subtracting 2 rocks. You are now back to the original number of rocks in the bucket.

You should also realize that the opposite of multiplication is division. If you multiply a number by 3, you can undo that by dividing by 3. They are opposite operations.

In the last lesson, you learned that the exponent 2 is read as "squared." For example, the equation below is read, "Four squared equals sixteen."

$$4^2 = 16$$

The opposite of squaring a number is the *square root*. Let me explain. You know that 4^2 (four squared) equals 16. But do you know the square root of 16? It's the opposite. The square root is the number we squared to get 16. Look at the equation below. It is read, "The square root of 16 equals 4."

$$\sqrt{16} = 4$$

Think of it this way, 4 is the *root*, when you *square* it, you get 16. You can undo that by getting the square root.

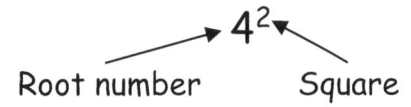

$$4^2$$

Root number   Square

Below, I have squared the number 3.

$$3^2 = 9$$

I can undo that by doing the opposite operation. The opposite is the square root of 9.

$$\sqrt{9} = 3$$

We are now back to our number before we squared it. We can also square a variable. Look at the example below.

$$x^2$$

To get x by itself, in the example above, do the opposite operation, the square root. The square root of x^2 is x.

$$\sqrt{x^2} = x$$

I can prove it to you. Let's pretend that x = 5 in the equation above. I will rewrite that problem with a 5 in place of the x.

$$\sqrt{5^2} = 5$$

Think about that for a moment. This means 5 x 5, which equals 25. And the square root of 25 is 5. Do you see how that works? Look at a few more examples.

$$\sqrt{4^2} = 4$$
$$\sqrt{8^2} = 8$$
$$\sqrt{6^2} = 6$$

Do you understand now? We squared a number and then we un-squared it with a square root symbol. So, if you ever end up with an answer that looks like the one below, you can easily undo a square by finding the square root.

$$x^2 = 5$$

To get x by itself, find the square root of x^2. And, of course, whatever you do on one side you must do to the other.

$$\sqrt{x^2} = \sqrt{5} \quad so \quad x = \sqrt{5}$$

Complete the next worksheet.

WORKSHEET 3-11

Solve the following.

1. $5^2 =$
2. $\sqrt{25} =$
3. $6^2 =$

4. $\sqrt{36} =$
5. $8^2 =$
6. $\sqrt{64} =$

7. $2^2 =$
8. $\sqrt{100} =$
9. $9^2 =$

10. $\sqrt{49} =$
11. $\sqrt{121} =$
12. $\sqrt{16} =$

13. $\sqrt{144} - \sqrt{16} =$
14. $\sqrt{4} + \sqrt{25} =$

15. $\sqrt{49} \cdot \sqrt{36} =$
16. $\sqrt{64} \cdot 2^2 =$

17. $\sqrt{100} \div 5 =$
18. $\sqrt{81} \cdot x = 36$

Name: _____   Date: _____

CHAPTER 1 REVIEW TEST

Solve for x.

1. $29 + x = 104$
2. $9x = 63$
3. $\frac{x}{4} = 6$

4. $\frac{48}{x} = 8$
5. $\frac{10}{x} \cdot \frac{14}{1} = 70$
6. $52 - x = 10$

7. $-201 - 3x = -225$
8. $\frac{1}{2}x + 12 = 10$
9. $\frac{11}{x} = 11$

10. $\frac{48}{x} + 12 = 24$
11. $\frac{9}{x} + 6 = 9$
12. $\frac{3}{5}x - 12 = -2\frac{2}{5}$

Solve for x in the following ratios.

13. $\frac{5}{x} = \frac{10}{12}$
14. $\frac{x}{32} = \frac{2}{8}$
15. $\frac{72}{9} = \frac{8}{x}$

Solve for x in terms of y.

16. $x + y = 325$
17. $\frac{x}{8} = y$
18. $x - 2 = y$

Solve the following.

19. $6^2 + 4^2 =$
20. $8^2 \cdot 10^2 =$
21. $\sqrt{81} \cdot \sqrt{36} =$

22. $5^3 \cdot y =$
23. $2^3 \div \frac{1}{6} =$
24. $\sqrt{100} \cdot 7^2 =$

CHAPTER 2

SOLVING ALGEBRAIC EXPRESSIONS

LESSON 12: TERMS, EXPRESSIONS AND EQUATIONS

Before we go any further, you need to learn the difference between a **term**, an **expression** and an **equation**. Below is a list of *Terms*. Each cluster of numbers/letters is one amount. They are not added or subtracted to any other amount. That's what makes them a term.

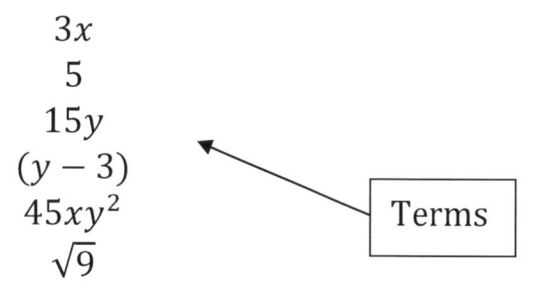

$$3x$$
$$5$$
$$15y$$
$$(y - 3)$$
$$45xy^2$$
$$\sqrt{9}$$

← Terms

An *Expression* is a set of terms "connected" with addition and/or subtraction. Below is a list of expressions.

$$3x + 2x + \sqrt{25}$$
$$(3 + 2) + x^2$$
$$18 - (1 + 5) + (10 \times 4)$$
$$3^3 + 4^2$$

← Expressions

An *Equation* has an equal sign in the middle and both sides are equal. Get it? Equa – tion...Equal.

$$6x = 36$$
$$5x + 2^2 = 14$$
$$(3 + 5) + (4 \times 5) = 7x$$

← Equations

Do you see the difference between a term, an expression and an equation? Look at each list of those lists carefully. A term is just one clump of numbers and/or variables. An expression combines two or more terms together with addition or subtraction. And an **equa**tion will always have an **equa**l sign with something on both sides.

Complete the next worksheet.

Name: _____   Date: _____

WORKSHEET 3-12

Name each of the following as a term, an expression, or an equation.

1. $3x$

2. $4x - 2x + 10$

3. $5y^2$

4. $5y + x = 40$

5. $-10 + 2y - 7y + 8$

6. $\sqrt{64}$

Find the value of each of the following terms.

7. $\sqrt{81} =$

8. $3^3 =$

9. If $x = 4$, then $3x =$

Find the value of the following expressions.

10. $3^2 + 7 =$

11. $\sqrt{36} - 2^2 =$

12. If $x = 5$, then $2x + 5x =$

Solve for x in the following equations.

13. $\sqrt{x} = 5$

14. $7x + 3 = 31$

15. $45 = 9x + 9$

Solve for m in the following equations.

16. $8m + 2 = 58$

17. $\frac{m}{8} = 8$

18. $m^2 = 7$

LESSON 13: COMBINING LIKE TERMS

In algebra, you will have to combine terms A LOT. I say "combine" because sometimes you will add, and other times you will subtract. This is called combining. You can only add or subtract terms if they are alike. Terms that are alike are called *Like Terms*.

For example, you can add 3x + 7x, but you cannot add 3x + 7y. The last two are not alike.

Think of it this way: let's pretend the x stands for cats and y stands for dogs. Try to add the following:

3 cats + 7 cats = 10 cats
3 dogs + 7 dogs = 10 dogs
3 cats + 7 dogs = WHAT? 10 catogs? No, they can't be added.

You can only add terms that are the same kind. They are called *like terms*. Look at the following expression.

$$2x + 6y + 4x + 8y$$

Do you see any like terms? 2x and 4x are like terms and so are 6y and 8y. Add the x's together and add the y's together. That's called adding like terms.

$$2x + 6y + 4x + 8y$$

$$2x + 4x = 6x \qquad 6y + 8y = 14y$$

$$6x + 14y$$

When we change one expression into another expression by combining like terms, that is called *simplifying*. We simplified the expression above by making it simpler. There are only two terms now instead of four.

If you are confused, read the last two pages again. It's not as complicated as you think. In that last problem, we just added up the x's and the y's.

Simplify this next expression by combining like terms. I separated each term, so it's easier to see what we are dealing with here.

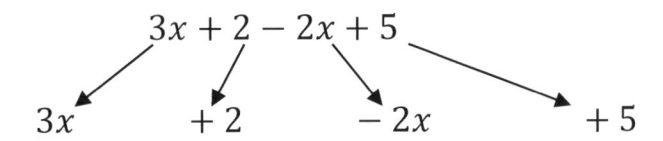

$$3x + 2 - 2x + 5$$

$$3x \qquad +2 \qquad -2x \qquad +5$$

Look for any like terms. 3x and -2x are like terms. When you look at each term, you also have to look at the sign in front of it. The 3x is positive, but the 2x is negative, so put them together like this and do the math.

$$3x - 2x = 1x$$

The terms that are left over, from the problem above, are +2 and +5. These are also like terms, so we can add them together, too.

$$2 + 5 = 7$$

Now that we have combined like terms, our original problem is simplified. We turned this, into this.

$$3x + 2 - 2x + 5 = 1x + 7$$

We can drop the number 1 in the answer because 1x is the same thing as just an x. Just like 1 cat is the same as a cat.

We have just simplified 3x + 2 - 2x + 5 by combining the x's and the non-x's. Now it is short and simple, x + 7.

Remember you can only add (or subtract) like terms. The variables and the exponents must be exactly the same to be like terms. Look at the list of terms that can and cannot be added together.

CAN	CANNOT
2x + 5x = 7x	2x + 5y
3xy + 7xy = 10xy	$3xy^2$ + 7xy
$5y^2 + 2y^2 = 7y^2$	$5y^2 + 2y^3$

Do you see how the slightest difference in the terms makes them not like terms? Look at this next problem.

$$4^2 + 4^2 =$$

To add these two terms together, solve each one and then add them together.

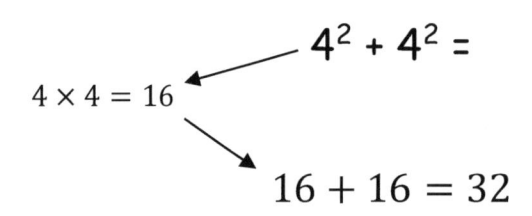

$$4^2 + 4^2 =$$
$$4 \times 4 = 16$$
$$16 + 16 = 32$$

That's easy to solve, so I'll make it a little more difficult. Look at this next problem.

$$x^2 + x^2$$

Since we don't know the value of x, we can't just add them together like we did with $4^2 + 4^2$. So, let me ask you this question. Is 16 + 16 the same thing as 16 x 2? Yes, it is...so it should be easy for you to believe that $x^2 + x^2$ is the same thing as $x^2 \cdot 2$. In algebra this is written as $2x^2$ because we like to see the number in front of the variables.

$$x^2 + x^2 = 2x^2$$

To make this even clearer, I will put an invisible number 1 in front of x^2.

$$1x^2 + 1x^2 = 2x^2$$

Do you see how this problem is basically 1 + 1 = 2? As long as the terms you are adding up are like terms, you can just add them together and put the new number in front.

Can you simplify this next problem?
$$x^3 + x^3 + x^3 =$$

How many x^3 do you see? I see 3, so the answer is $3x^3$.

We will look at another one.
$$3a^4 + 2a^4 =$$

Can you guess how to add these two terms together? Look at the two different numbers of a^4. They are 3 and 2. Think of a^4 as…umm…a type of apple! Now the problem is read as, "3 apples plus 2 apples," which of course equals 5 apples.

$$3a^4 + 2a^4 = 5a^4$$

However, we cannot add apples and oranges. Look at this next addition problem.
$$x^3 + a^4 =$$

We cannot add these two terms together. If "x^3" stands for oranges and "a^4" stands for apples, how can you add these together? Oranpples? No, that's not right. You cannot add these two together because they are not like terms.

The rule to be learned here is that you can only add *like terms*. And they must be exactly alike too. The numbers in front don't need to be the same, just the exponents and variables. You cannot add these two terms either.
$$5a^2 + 5a^3 =$$

These cannot be added together because they are not like terms. An a^2 apple is completely different than an a^3 apple; everyone knows that, so you cannot add those terms together.

Subtraction of exponents works the same way. You can only subtract like terms. Can you solve this one?

$$4xy^2 - 2xy^2 =$$

Since these two terms are alike, we can just subtract 4 - 2 = 2, so the answer is $2xy^2$.

Can you subtract these terms?

$$7m^3 - 4m^2 =$$

No, you cannot. These are not like terms. You cannot simplify this problem any further. To *simplify* a problem means to do as much math as you can to make it smaller, simpler. Try some problems on your own by completing the next worksheet.

If you understand everything in this book so far, complete the next worksheet. If you are confused, go back to the lesson you didn't understand. And realize that you are learning several different rules about algebra. They might not (in fact, I'm sure they don't) make sense to you right now, but soon all these little rules will come together, so you can solve difficult algebra problems.

Name: _____   Date: _____

WORKSHEET 3-13

Simplify each of the following expressions by combining like terms.

1. $3y + 4y - 2x$

2. $4x - 2x + 3y - 2y$

3. $7y - 2x + 4y$

4. $6x - 5x + 8y + 2x$

5. $8m - 2m + 3r$

6. $7x + 7y + 7m + 3x$

7. $4x^2 + 3x^2$

8. $8rst + 9rst + rst$

9. $5 + 7b^2 - 2 - 3b^2$

10. $5xy + 7xy + 17xy^2$

11. $3mn - mn$

12. $6xy^2 + 5xy^2 + 8xy^3$

13. $18y + 2$

14. $27rst + 33rst - 7rst + 4rs$

LESSON 14: MULTIPLYING TERMS

If you feel confident with adding and subtracting terms with exponents, then you are ready to learn how to multiply and divide with them, too. Take a look at this multiplication problem.

$$y^2 \cdot y^4 =$$

What does this problem mean? The exponents tell us how many times to use "y" in a multiplication problem.

$$y^2 \cdot y^4 =$$

(twice) $y \cdot y$ · (four times) $y \cdot y \cdot y \cdot y$

Squish this problem together and you get $y \cdot y \cdot y \cdot y \cdot y \cdot y$. Multiplying "y" by itself 6 times is the same thing as y^6. So, it is easy to get the answer. Just add the exponents together.

$$y^2 \cdot y^4 = y^{2+4} \quad \text{or } y^6$$

Let's try another one.

$$x^2 \cdot x^5 \cdot x^4 =$$

Since the bases are the same, just add up the exponents $2 + 5 + 4 = 11$.

$$x^2 \cdot x^5 \cdot x^4 = x^{11}$$

When the bases are not the same, as in this next problem, it's a little different, but just as easy.

$$x^2 \cdot y^2 =$$

You can't add up the exponents when the bases are different because x and y represent two different numbers. Instead, we just squish them together. When two letters are next to each other, it already means they are being multiplied, so just remove the dot. Put the letters in alphabetical order and squish them together.

$$x^2 \cdot y^2 = x^2y^2$$

The reason we arrange the variables in alphabetical order is, so we all end up with x^2y^2. You could switch them around to be y^2x^2, it means the same thing, but algebra is confusing enough. Let's not complicate things by having different answers.

Another thing you should realize is that when a number or letter doesn't have an exponent, it is considered to have an exponent of 1; an invisible 1. For example, look at this next problem.

$$a \cdot a^2 =$$

That is the same thing as $a^1 \cdot a^2$, so just add up the exponents and the answer is a^3. Or simply look at that problem and ask yourself, "How many a's are being multiplied together?" I see 3. That is the exponent for the answer.

Now let's mix it up a little bit. I will throw in some terms with different bases. See if you can simplify this problem.

$$y^2 \cdot y^4 \cdot x^2y$$

How many y's are being multiplied? How many x's are being multiplied? The first term has 2 y's. The second term has 4 y's, and the third term has 1 y and 2 x's. Squish them all together, in alphabetical order, with the new exponents.

$$y^2 \cdot y^4 \cdot x^2y = x^2y^7$$

Now let's throw in some numbers, too. Just multiply the numbers as you normally would. Then tally up the exponents and put them above the appropriate base.

$$3x^2y^2 \cdot 4xy =$$

$$3x^2y^2 \cdot 4xy =$$

Can you solve this one in your head?
Multiply the numbers first, 3 x 4 = 12.
How many x's are being multiplied? 3
How many y's are being multiplied? 3
The answer is $12x^3y^3$.
Here are the two algebra skills you just learned:

- To multiply similar bases together, add the exponents $2x^2 \cdot 3x^3 = 6x^5$
- To add like terms together, add up the whole numbers. $2x + 3x = 5x$

Use these skills to complete the next worksheet.

Name: _____   Date: _____

WORKSHEET 3-14

1. $x^2 \cdot x^2 =$

2. $x^3 \cdot x^4 =$

3. $y^6 \cdot y^5 =$

4. $xy^2 \cdot xy =$

5. $abc^2 \cdot ab \cdot abc =$

6. $xyz \cdot x^4 y^3 z =$

7. $4x^2 \cdot 3x^3 =$

8. $7ab^3 \cdot ab^3 =$

9. $9a \cdot 3a =$

10. $7x \cdot x^3 =$

11. $x \cdot y =$

12. $5x^3 \cdot 7y^4 =$

13. $11y^2 \cdot 11xy^2 =$

14. $x^{10} \cdot x^{10} =$

15. $9abc^2 \cdot 3xyz^2 =$

16. $14a^2 \cdot a =$

17. $x \cdot y \cdot z \cdot z =$

18. $a^2 b^2 c^2 \cdot abc =$

19. $\frac{3}{5} x \cdot x =$

20. $8^2 \cdot y^2 =$

21. $21x^2 \cdot x^2 + 3y =$

LESSON 15: MATH INSIDE OF PARENTHESES

Sometimes, you will find an equation that has a problem inside of parentheses.

$$(2 + 3) + x = 10$$

When you see parentheses, you MUST do that math first.

$$(2 + 3) + x = 10$$
$$\downarrow$$
$$5 + x = 10$$

Now we are ready to solve for x like we did before.

$$5 + x = 10$$
$$x = 10 - 5$$
$$x = 5$$

Let's try a few more together. Solve for x. Start with the math inside the parentheses.

$$x + (3 \times 2) = 20$$
$$\downarrow$$
$$x + 6 = 20$$
$$x = 20 - 6$$
$$x = 14$$

Here is another one. Follow along and make sure you understand each step.

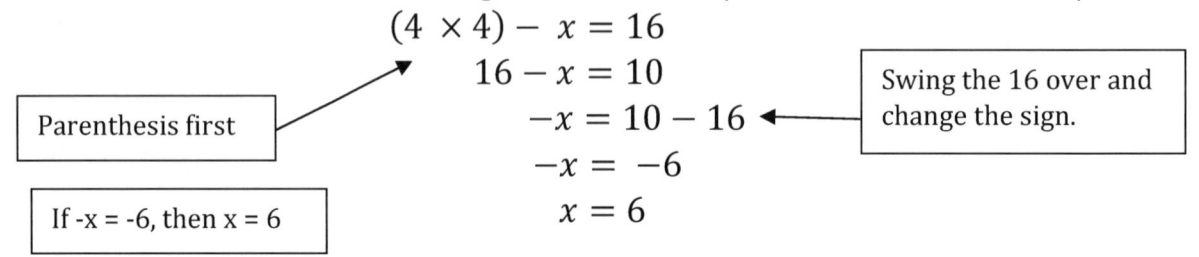

$$(4 \times 4) - x = 16$$
$$16 - x = 10$$
$$-x = 10 - 16$$
$$-x = -6$$
$$x = 6$$

Parenthesis first

If -x = -6, then x = 6

Swing the 16 over and change the sign.

Read through this one, too. Make sure you understand each step.

$$3x + (2 + 2) = 16$$
$$3x + 4 = 16$$
$$3x = 16 - 4$$
$$3x = 12$$
$$\frac{3x}{3} = \frac{12}{3}$$
$$x = 4$$

Sometimes there will be parentheses inside of parentheses or brackets. When this happens, start from the inner most set of brackets.

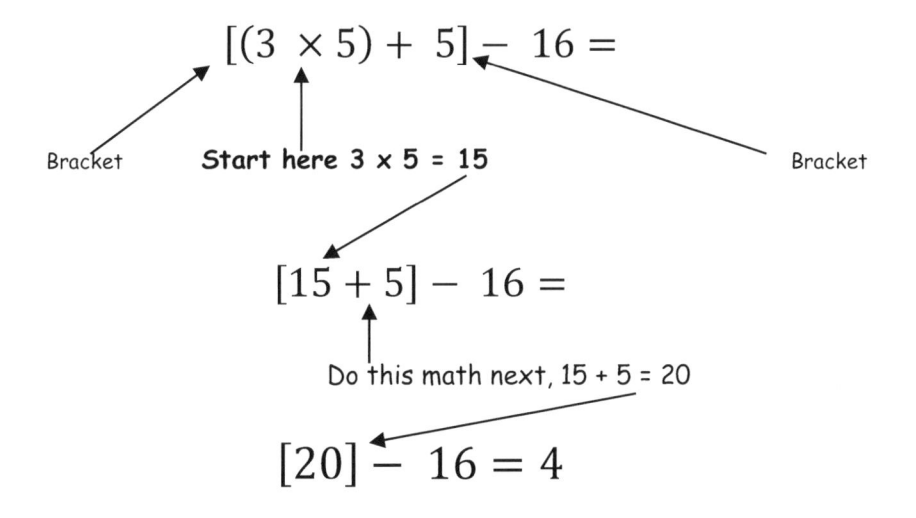

$$[(3 \times 5) + 5] - 16 =$$

Bracket   **Start here 3 x 5 = 15**   Bracket

$$[15 + 5] - 16 =$$

Do this math next, 15 + 5 = 20

$$[20] - 16 = 4$$

Always remember to start with the parenthesis. The inner most set first.

If you understand everything in this book so far, complete the next worksheet. If your eyebrows are down because you don't fully get it, go back to the beginning. It won't take long to read it a second time.

Name: _____   Date: _____

WORKSHEET 3-15

Solve for x.

1. $25x = (63 + 37)$

2. $52 + x = (33 \cdot 3)$

3. $x + (13 - 4) = 15$

4. $(7 \cdot 3) + x = 126$

5. $x + (5 \cdot 5) = 45$

6. $6x = (14 + 16) \cdot 2$

7. $(15 \div 3) + x^2 = 54$

8. $\sqrt{25} + (3 \cdot 5) = 2x$

9. $\sqrt{x} + (2^2 \cdot 3^2) = 45$

10. $(x \cdot 3) + 2 = 29$

11. $(12 \div 4) \cdot 2x = 54$

12. $(-5 \cdot -5) \div x = 5$

13. $\left(\frac{1}{4} + \frac{12}{16}\right) \cdot 2x + 3 = 5$

14. $(x \cdot x) + 4 = 53$

15. $[(12 + 14) \div 2] + 2x = 27$

16. $[(7 \cdot 3) + 4] - 5 = 4x$

LESSON 16: THE SECOND OPERATION

In the beginning of this book, you learned the two main rules to solving an algebraic equation. They are "get x by itself" and "whatever you do to one side of the equal sign you must do to the other side." That is still true, but as the equations get more difficult it is going to take more math to get x by itself.

Eventually, the equations will get so intense it is hard to know where to start. That's why mathematicians all agree to go in the same order. That will also ensure that everyone gets the same answer because it is possible to get two different answers if you don't follow the Proper Order of Operations. And only one answer is right.

In the last lesson, you learned the very first thing to look for in an algebraic equation. You are to look for, and solve, any math inside of parentheses. If there are parentheses or brackets inside of parentheses, you are to start with the inner most set. If there are no parentheses or brackets in the equation you are trying to solve, then skip it and move on to the second operation.

The second procedure to follow in the Proper Order of Operations is solving any exponents or square roots. Let's take a look at an equation with both parentheses and exponents in it and solve it together.

Parenthesis first
$$7^2 + (5 + 3) - x = 47$$

$$7^2 + (8) - x = 47$$
Exponents and square roots are always solved second.

$$49 + 8 - x = 47$$
Add 49 + 8 and bring down the rest of the equation.

$$57 - x = 47$$

Subtract 57 from both sides to get x by itself.

$$-x = -10$$

If -x = -10, then x = 10

$$x = 10$$

If your equation has both an exponent and a square root sign, you are to solve them in the order they appear when reading from left to right. I'll show you what I mean. Where do we start in the equation below?

$$\sqrt{25} + 8^2 + x - (5 + 2) = 88$$

According to the Proper Order of Operations, we always start with the math in the parentheses. Solve that and then bring down the rest of the equation.

$$\sqrt{25} + 8^2 + x - (5 + 2) = 88$$
$$\downarrow$$
$$\sqrt{25} + 8^2 + x - 7 = 88$$

Next up, we are to look for any exponents and square root signs. I see both in the equation above. According to the Proper Order of Operations, we read the problem from left to right and then solve whichever one comes first.

$$\sqrt{25} + 8^2 + x - 7 = 88$$

The first term in our problem is the square root of 25. Do you remember what that is? The square root of 25 is 5 because 5 x 5 = 25. Next, we have to find the value of 8^2. That is the same thing as 8 x 8, so the answer is 64. I'll rewrite the equation with the work we've done, so far.

$$5 + 64 + x - 7 = 88$$

Now we can combine the whole numbers and then solve for x like we did before.

$$5 + 64 + x - 7 = 88$$

$$5 + 64 - 7 = 62$$

Now our problem looks like this:

$$x + 62 = 88$$

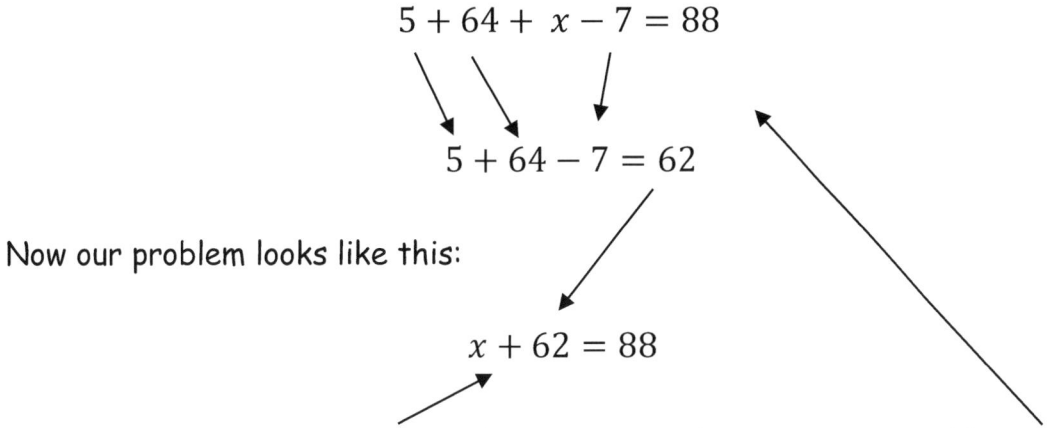

Can you figure out why this is a plus sign now and not a minus sign like it was up there? I almost wrote $x - 62 = 88$, but that's not right. When we combined all the whole numbers, we didn't do anything with the x. It just stayed there as a positive variable. The answer of our combined whole numbers was positive 62, so that's what we had to put next to our x. The results are below.

$$x + 62 = 88$$

The only thing left to do is swing the 62 over to the other side and change the sign.

$$x = 88 - 62$$

$$x = 26$$

Here is the entire problem written out the way it should look on your paper.

$$\sqrt{25} + 8^2 + x - (5 + 2) = 88$$
$$\sqrt{25} + 8^2 + x - (7) = 88$$
$$5 + 64 + x - 7 = 88$$
$$69 + x - 7 = 88$$
$$62 + x = 88$$
$$x = 88 - 62$$
$$x = 26$$

Here is a tricky one.

$$4^3 \cdot 3x = 1{,}344$$

There are no parentheses, so we look for any exponents or square roots. There is one exponent, so solve 4^3 and write it down below.

$$4 \times 4 \times 4 = 64 \quad \begin{array}{l} 4^3 \cdot 3x = 1{,}344 \\ 64 \cdot 3x = 1{,}344 \end{array}$$

We can't subtract 64 because it is part of a multiplication problem.

We need to multiply $64 \cdot 3x$ before we can move on.

$$64 \cdot 3x = 1{,}344$$

Look at $64 \cdot 3x$ and realize that x is a number, too. Solving $64 \cdot 3x$ is the same thing as:

$$64 \cdot 3 \cdot x = 1{,}344$$
$$192 \cdot x = 1{,}344$$
$$192x = 1{,}344$$

Divide both sides by 192 to get x by itself.

$$\frac{192x}{192} = \frac{1{,}344}{192}$$

$$x = 7$$

Did I get it right? Put 7 in place of the x in the original problem and see if it makes sense.

$$4^3 \cdot 3x = 1{,}344$$
$$64 \cdot 3 \cdot 7 = 1{,}344$$

Yep, it's right.

This next problem has a square root to solve. The correct order is parentheses and then exponents/square roots. Start by solving the math inside the parentheses and then read the problem from left to right. Solve any exponents or square roots in the order they appear.

Find the value of this next expression. To "find the value," means to find the answer.

$$7 + 8 - 3^2 + (8 \cdot 7) - \sqrt{25} + 4$$

Do you know where to start? That's right, start with the parentheses.

$$7 + 8 - 3^2 + (8 \cdot 7) - \sqrt{25} + 4 =$$
$$7 + 8 - 3^2 + (56) - \sqrt{25} + 4 =$$

What do you do next? Read from left to right and solve any exponents or square roots in the order they appear.

$$7 + 8 - 3^2 + (56) - \sqrt{25} + 4 =$$

The first one we run into is 3^2. Next is $\sqrt{25}$. Solve those two and then rewrite our problem. You can drop the parentheses, too.

$$7 + 8 - 9 + 56 - 5 + 4 =$$

We are left with an addition and subtraction problem. I have written out the math, step by step.

$$7 + 8 - 9 + 56 - 5 + 4 =$$

$7 + 8 = 15 \longrightarrow 15 - 9 = 6 \longrightarrow 6 + 56 = 62 \longrightarrow 62 - 5 = 57 \longrightarrow 57 + 4 = 61$

The final answer, or the value of the expression, is 61.

$$7 + 8 - 3^2 + (8 \cdot 7) - \sqrt{25} + 4 = 61$$

Complete the next worksheet. If this worksheet is difficult for you, go back a few lessons until the math is easy. You must understand each new skill before you can continue.

Name: _____   Date: _____

WORKSHEET 3-16

Solve for x in the equations below. Be sure to go in the proper order.

1. $2^2 - x + (3 \cdot 2) = 8$

2. $(4 \cdot 5) - 4^2 + x = 24$

3. $3^2 + (8 \cdot 3) = 3x$

4. $7x + 5x - (4 \cdot 2x) = 10^2$

5. $2^3 - (6 \cdot x) = -4$

6. $\frac{(9^2 + 3^2)}{x} = 9$

Find the value (the answer) for the expressions below.

7. $\sqrt{9} + 2^2 - [20 - (6 \cdot 3)] =$

8. $(7 - 5) \cdot 3^2 \cdot \sqrt{4} =$

9. $\left(\frac{1}{2} + \frac{3}{8}\right) \cdot \left(\frac{2}{3} - \frac{1}{6}\right) \cdot 2^2 =$

10. $\frac{\sqrt{100} \cdot 6^2}{2} =$

LESSON 17: P.E.M.D.A.S

You already know that the first thing to do when finding the value of an expression is to start with the parentheses. The next operation is to solve any exponential terms or square roots, in order from left to right.
Now we are ready for the third operation. It is multiplication and division. Look at the title above, P.E.M.D.A.S. This stands for:

Parentheses
Exponents and square roots
Multiplication
Division
Addition
Subtraction

This is the order you follow when you are finding the value of an expression. It is called the *Order of Operations*. Many people like to remember this order by remembering this phrase.

Please Excuse My Dear Aunt Sally.

If you don't like that phrase, you can always come up with your own phrase that has the initials P.E.M.D.A.S.

The reason we must follow this order is, so everyone gets the same answer. Look at the expression below. It is possible to get two different answers if you don't follow the order of operations.

$$3 + 7 \times 3 =$$

What does this problem mean?

$$3 + 7 \times 3 =$$

$$10 \times 3 = 30 \quad or \quad 3 + 21 = 24$$

Starting with addition gives you this answer. Starting with multiplication gives you the second answer. How do you know which one is right? That's why mathematicians have agreed to follow the order of operations, to eliminate this confusion.

Let's find the value of the following expression by using the PEMDAS order of operations.

$$48 \div (13 - 7) + 2^3 \times (6 - 2) =$$

First on the list is Parentheses. Solve those two problems first and then bring down the rest of the expression.

$$48 \div (13 - 7) + 2^3 \times (6 - 2) =$$
$$48 \div (6) + 2^3 \times (4) =$$

Next, we solve any exponents. 2^3 means 2 x 2 x 2 = 8.

$$48 \div (13 - 7) + 2^3 \times (6 - 2) =$$
$$48 \div (6) + 2^3 \times (4) =$$
$$48 \div (6) + 8 \times (4) =$$

We can drop those parentheses. We don't need them anymore.

$$48 \div 6 + 8 \times 4 =$$

The next two operations are multiplication and division. These two go together. Starting from the left side, solve any multiplication or division in the order they appear. In this problem, division is first because we read from left to right. If multiplication had come first, you would solve that. Just read from left to right and solve each multiplication or division problem as they show up.

$$48 \div 6 + 8 \times 4 =$$
$$8 + 32 = 40$$

The last two operations are addition and subtraction. They are also solved in order from left to right. Our problem above only has addition. Solve that math and you are finished.

Look again at the Order of Operations:

Parentheses ⟶ Always first
Exponents and square roots ⟶ Opposite operations
Multiplication
Division ⟶ Opposite operations
Addition
Subtraction ⟶ Opposite operations

Notice that there are actually only 4 steps above because when you get to an opposite set of operations, you do those two at the same time. It might be easier to remember P.E.M.A.

    Parentheses, inner most first
    Exponents and the opposite operation, square root
    Multiplication and division
    Addition and subtraction

Let's try a few more. Find the value of this expression. Don't be intimidated by the looks of this expression - it's easier than you think.

Brackets

$$4^3 - \sqrt{16} \div [(3 \cdot 8) - 22] =$$

Where do you start?

Start with the parenthesis inside the brackets.

$$4^3 - \sqrt{16} \div [(3 \cdot 8) - 22] =$$

$$4^3 - \sqrt{16} \div [(24) - 22] =$$

Next, solve the math inside the brackets.

$$4^3 - \sqrt{16} \div [2] =$$

That takes care of the "P" in PEMA. Next are Exponents and square roots, as they appear from left to right.

$$4^3 - \sqrt{16} \div [2] =$$

$$64 - 4 \div 2 =$$

That takes care of the "PE" in PEMA. The next two operations are Multiplication and Division, as they appear from left to right.

$$64 - 4 \div 2 =$$

There is no multiplication in this problem, so move on to division.

$$64 - 4 \div 2 =$$

$$64 - 2 = 62$$

The last steps are addition and subtraction. This problem is finished.

Here is another one. Find the answer to this next expression.

$$7 \times 8 + (4 \times 10) - 8^2 =$$

Start with the parenthesis.

Parentheses: $7 \times 8 + (40) - 8^2 =$

Exponents/Square root: $7 \times 8 + (40) - 64 =$

Multiplication/Division: $56 + 40 - 64 =$

Addition/Subtraction: $96 - 64 =$

Final answer: $96 - 64 = 32$

Below is an expression. Solve it to find the value.

$$12 + 5^2 \cdot 4 \div \sqrt{4} =$$
P. (There aren't any)

$$12 + 5^2 \cdot 4 \div \sqrt{4} =$$
E. and S.R. (left to right)

$$12 + 25 \cdot 4 \div 2 =$$
M. and then D. (left to right)

$$12 + 100 \div 2 =$$

$$12 + 50 =$$
A. and S.

$$12 + 50 = 62$$

Try a few on your own. Make sure you follow the order of operations, when finding the value of an expression. If you have any difficulties, read this lesson again.

Name: _____   Date: _____

WORKSHEET 3-17

Find the value of the following expressions.

1. $\sqrt{81} \cdot 4 + (\sqrt{4} + \sqrt{16}) =$

2. $10^2 - 2 \cdot 5^2 + 2x =$

3. $3x + 2x - [(6^2 + 4) \cdot 2] =$

4. $4 \cdot (3 + 5) - \sqrt{49} \cdot \sqrt{4} =$

Solve for x in the following equations by getting x by itself.

5. $3x + 7^2 - \sqrt{25} = 74$

6. $56 - 5x = 11$

7. $3^2 - x + (3 \cdot 2) = 11$

8. $(x \cdot x) + 5^2 = 125$

Simplify the following expressions by combining like terms.

9. $3x - 7x + x + 2y$

10. $4x^2 \cdot 5x^2 + 2x + 6x$

11. $6y \cdot 5y^2 - 6x + 7y^3$

12. $3y^3 \cdot 2y^3 + 5y^6$

LESSON 18: DISTRIBUTIVE PROPERTY OF MULTIPLICATION

Look at the math below.

$$3 \times (4 + 2) =$$

You can look at this two different ways. One way is to solve the math inside the parentheses and then multiply.

$$3 \times (4 + 2) =$$
$$3 \times (6) =$$
$$= 18$$

The other way is to use the *Distributive Property of Multiplication*. This little law of algebra gives us another way to solve this problem. Let's solve the same problem using this new law.

The Distributive Property of Multiplication says that we can multiply the number on the outside of the parentheses, by each of the numbers on the inside. Then, we add those two answers together to get the final answer.

$$3 \times (4 + 2) = \qquad 3 \times (4 + 2) =$$
$$12 \qquad + \qquad 6 \qquad = \qquad 18$$

It still equals 18. That is how the Distributive Property works. You can multiply the numbers separately and then add the answers together. Or you can add them before you multiply; either way you get the same answer.

Let's try another one. Only this time, we will drop the multiplication sign on the outside of the parentheses. From now on, if you see a number next to a set of parentheses, like this next example, know that it means multiplication.

$$6(8 + 6) \text{ means } 6 \times (8 + 6)$$

You can solve this problem two different ways. One way is to add the numbers inside the parentheses and then multiply that by 6.

$$6(8 + 6) = 6 \times 14 = 84$$

The other way is to multiply the number on the outside by each of the numbers on the inside and then add those answers together.

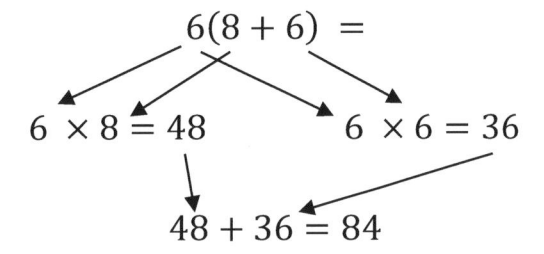

$$6(8 + 6) =$$

$$6 \times 8 = 48 \qquad 6 \times 6 = 36$$

$$48 + 36 = 84$$

The second method may seem longer, but sometimes it is your only choice. Look at this next problem. We can't just add up the math inside the parentheses, so we will have to use the law of distribution on this one.

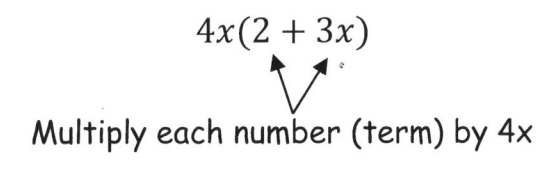

$$4x(2 + 3x)$$

Multiply each number (term) by 4x

$$(4x \cdot 2) + (4x \cdot 3x) =$$

$$8x + 12x^2$$

These two terms are not like terms, so this problem is finished.

Let's try one more. First, we will multiply the outside number by the first term in the parentheses. Then, we will multiply the outside number by the second term in the parentheses. Add your answers together. I've written out the math for you. Read through it until it makes sense.

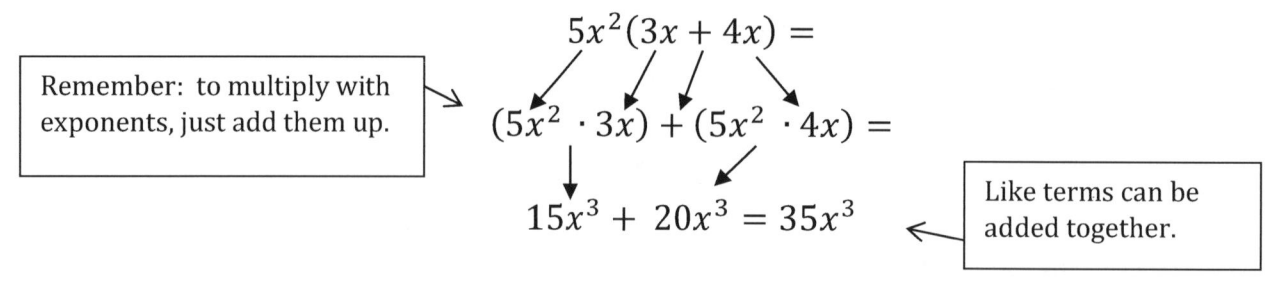

Remember: to multiply with exponents, just add them up.

$$5x^2(3x + 4x) =$$
$$(5x^2 \cdot 3x) + (5x^2 \cdot 4x) =$$
$$15x^3 + 20x^3 = 35x^3$$

Like terms can be added together.

These two numbers are like terms, so we can add them together. The answer is $35x^3$.

Now that you have seen how the math is done, we can skip a step by solving some of the math in our mind. I'll show you what I mean in the next example.

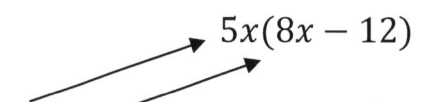

$$5x(8x - 12)$$

Can you multiply $5x \times 8x$ in your mind? I can! First, I'll multiply the coefficients, $5 \times 8 = 40$. Next, I'll multiply the two variables. The answer is $40x^2$. I'll write that below the problem.

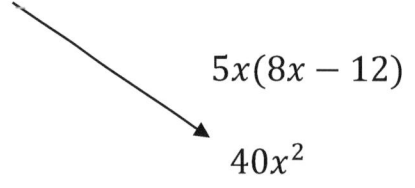

$$5x(8x - 12)$$
$$40x^2$$

Next, we need to multiply $5x \times -12$. That is the tricky part! You must include the sign when you multiply in your mind. We have to multiply by NEGATIVE 12. The answer is $-60x$. I'll put that next to our last answer.

$$5x(8x - 12)$$
$$40x^2 - 60x$$

That's it, the problem is solved. Do you see how we skipped the step with the two sets of parentheses? This way is much faster and cleaner. Let's try one more together using this faster method.

$$-4x(7xy - 6x) =$$

Try to solve this one on your own before you read my solution.

What did you get when you multiplied the first two terms? I got $-28x^2y$. That is the answer to $-4x \times 7xy$. What did you get for the second term? Did you remember to multiply with negative numbers? The second part is $-4x \times -6x = 24x^2$. The entire answer is $-28x^2y + 24x^2$. These two terms are not like terms, so we cannot combine them (add them together).

If you got the same answer that I did, then complete the next worksheet. If you got the wrong answer, read this lesson again.

Name: _____ Date: _____

WORKSHEET 3-18

Solve the following by using the distributive property of multiplication.

1. $2(6+9) =$

2. $3x(2x-3) =$

3. $5x^2(2x+5x^2) =$

4. $20x^3(3x-7x^2) =$

5. $3x(y-x) =$

6. $9^2(x+y) =$

7. $-4(2x-3) =$

8. $-3a^2(a+b) =$

9. $m(x+y) =$

10. $5rs(6rs-rs) =$

11. $ab(3a^2b-4ab^2) =$

12. $\frac{1}{2}(a+b) =$

13. $-5x(x^3-2x^2) =$

14. $-4a(3a-2b) =$

LESSON 19: SOLVING ALGEBRAIC EQUATIONS

Let's review all the new "algebra moves" you have learned, so far. You know how to solve for x in the problem below, right?

$$3x = 15$$

Of course, you do. Just divide both sides by 3 and you'll get $x = 5$. And you know how to solve for x in terms of y.

$$x + 7 = y$$

Just swing the 7 over to the other side, change the sign, and you are finished. When you solve for x in terms of y, the letter y will always be in the answer somewhere. In this case, the answer is $x = y - 7$.

You learned how to add terms with exponents. Like the ones below.

$$2a^3 + 4a^3 = 6a^3$$

As long as the variables and exponents are exactly the same in both terms, you can just add the two numbers in front. The numbers in front are called the *coefficients*. In the three terms above, the numbers 2, 4, and 6 are the coefficients in each term.

You also learned how to multiply terms with exponents. That one is easy; just add the exponents together and multiply the coefficients.

$$4b^4 \times 5ab^4 = 20ab^8$$

Remember, you can multiply any two terms together, but when it comes to adding or subtracting terms, they MUST be ALIKE, in other words, *like terms*.

Below are two equations. One can be added and the other one cannot. Can you tell which one can be solved?

$$7efg^2 + 8efg = \qquad 6c^2d^2 + 2c^2d^2 =$$

The problem on the right can be solved. The terms on the left are not like terms.

Do you remember the proper order of operations? They are: Parentheses, Exponents, Multiplication, Division, Addition, and Subtraction. The initials spell the word PEMDAS. Some people use the phrase, "Please Excuse My Dear Aunt Sally" to help them remember the proper order.

We all agree to solve equations in this order, so everyone gets the same answer. We also agree to write terms with the coefficient in front and then list all the variables in alphabetical order. That way everyone ends up with the same answer. In the example below, you must multiply first and then add $3ab^2c$ to that answer. Notice how I wrote the answer in alphabetical order with the coefficient in front.

$$3ab^2c + 8b \cdot 4abc = 35ab^2c$$

In the previous lesson, you learned how to use the Distributive Property of Multiplication. Can you solve this one?

$$3a(2a + 3ab) =$$

Since the terms inside the parentheses are not like terms, we can't add them together. The only thing we can do is *distribute the multiplication*. You see, when you look at an algebraic problem, you always look to see if there is any math that can be done. In the problem above, we can multiply. So, let's multiply and then see if we can do anything else.

$$3a(2a + 3ab) =$$
$$6a^2 + 9a^2b$$

We are left with two terms that are NOT like terms, so there is nothing else we can do here. The problem is simplified as far as possible.

Let's try another one.

$$x^2(x + 3x^3) =$$

What can we do? Can we add the terms in the parentheses? No, they are not like terms. So, what can we do? We can multiply by using the distributive property (or law) of multiplication. I've done the math below, but I'm sure you could have solved it in your mind.

$$(x^2 \cdot x) + (x^2 \cdot 3x^3) =$$

$$x^3 + 3x^5$$

Can you combine these terms? No, they are not like terms. This problem is complete.

Here is another long algebraic equation. Can you solve for x?

$$3x + \sqrt{16} = 5x - 3^2$$

Where do you start? There aren't any parentheses, so according to PEMDAS, we should solve any exponential terms or square root signs.

$$3x + 4 = 5x - 9$$

Now we want to get x by itself. I'll move the 4 over to the other side.

$$3x = 5x - 9 - 4$$

Next, I will move the 5x over to the other side.

$$3x - 5x = -9 - 4$$

Both sides of the equation have like terms, so let's combine them.

$$-2x = -13$$

We still don't have x by itself, so let's divide both sides by -2.

$$\frac{-2x}{-2} = \frac{-13}{-2}$$

$$x = \frac{-13}{-2}$$

That is a division problem (it's also an improper fraction). I will divide negative 13 by negative 2.

$$x = 6\frac{1}{2}$$

Practice using all your new algebra moves on the next worksheet.

Name: _____   Date: _____

WORKSHEET 3-19

Simplify the following expressions.

1. $2(5 - x) =$

2. $3x(6x + 4) =$

3. $2x(2x + 3) =$

4. $-x(y - 2) =$

Solve for x in the following equations.

5. $4x + 2x = 48$

6. $3^3 - 4x + (6 \cdot 5) = 41$

7. $3x(8 + 3) - (4^2) = 83$

8. $\sqrt{x} = (7 \cdot 2^2) - (7 \cdot \sqrt{9})$

Find the value for each of the following 3 expressions. Be sure to follow the correct order of operations, PEMA.

9. $(48 - 4) \div 2^2 + 2 \cdot 2^3 =$

10. $(-20 + 4) \div 2^3 - 3 \cdot 4 =$

Solve for x in terms of y.

11. $7 - x = y$

12. $x + y = 20$

13. $x \div 5 = y$

14. $\frac{36}{x} = y$

LESSON 19A: SOLVING EXPRESSIONS WITH THE DISTRIBUTIVE PROPERTY OF MULTIPLICATION

In that last lesson, you used the Distributive Property of Multiplication to solve algebra problems like the one below.

$$3x(4 + 16x)$$

In this lesson, you will do the same thing, but the problems will be much longer. Here is the first example.

$$3x(4 + 16x) - 8x(11 - 8x) =$$

Focus your attention on just the two terms in bold.

$$\mathbf{3x(4} + 16x) - 8x(11 - 8x) =$$

I will multiply just those two terms together and then write the answer below.

$$\mathbf{3x(4} + 16x) - 8x(11 - 8x) =$$
$$12x$$

Now look at just these two terms. Don't forget their signs.

$$\mathbf{3x(4 + 16x)} - 8x(11 - 8x) =$$

I will multiply those two terms and add that answer below; $3x \times 16x = 48x^2$

$$\mathbf{3x(4 + 16x)} - 8x(11 - 8x) =$$

$$12x + 48x^2$$

I put a plus sign here because the answer to $3x \times 16x$ is a positive answer.

Now look at the next two terms for us to multiply. Don't forget the negative sign when you do the math.

$$3x(4 + 16x) - 8x(11 - 8x) =$$

$$12x + 48x^2 - 88x$$

I multiplied $-8x \times 11$ and put the answer here. Let's multiply the final two terms.

$$3x(4 + 16x) - 8x(11 - 8x) =$$

$$12x + 48x^2 - 88x + 64x^2 =$$

The only thing left to do is combine any like terms. Try it on your own and then see if we both have the same answer.

I combined $12x$ and $-88x$ to get $-76x$. Then I combined the two x² terms and got $112x^2$. I'll put the two answers together to get $-76x + 112x^2$. Is that what you got? It's OK if you wrote $112x^2 - 76x$. It means the same thing. Let's try one more together.

First up is a 9x. Is it negative or positive? Since it doesn't have a sign, it is positive.

$$9x(3x - 4y) - 6x(3y - 4x)$$

We are going to multiply that 9x by 3x. Is the 3x negative or positive? Since it doesn't have a sign, it is positive too. Those are the first two terms to be multiplied. I have written the answer below.

$$9x(3x - 4y) - 6x(3y - 4x)$$
$$27x^2$$

Next, we need to multiply positive 9x by negative 4y. The answer is below.

$$9x(3x - 4y) - 6x(3y - 4x)$$
$$27x^2 - 36xy$$

Do you know what to multiply next? That's right, $-6x \times 3y = -18xy$.

$$9x(3x - 4y) - 6x(3y - 4x)$$
$$27x^2 - 36xy - 18xy$$

There are only two terms left to multiply. I can solve that part in my mind. Both terms are negative, so I know the answer will be positive. I'll just multiply 6 x 4 and square the x's.

$$9x(3x - 4y) - 6x(3y - 4x)$$
$$27x^2 - 36xy - 18xy + 24x^2$$

The last thing to do is combine the like terms. The final answer is below.

$$51x^2 - 54xy$$

Give it try yourself by completing the next worksheet.

Name _____ Date _____

WORKSHEET 3-19A

1. $2x(3y - 4x) + 3y(6x + 9y) =$ 2. $5a(7a - 6b) - 3b(4a - 2b) =$

3. $-5x(2x - 8y) - 7x(4x + 9y) =$ 4. $-x(x + y) + y(y - x) =$

5. $-.35a(-1.2a + 3b) - .14a(3.1b - 6.2a) =$

6. $\frac{3}{8}x\left(\frac{1}{2}y - \frac{1}{4}x\right) - \frac{7}{16}x\left(\frac{5}{8}y + \frac{1}{8}x^2\right) =$

115

WORKSHEET 19A page 2

7. $14mn(mn + 8n) - 8m(4n - 6n^2) =$

8. $.5x(2xy - 4.2y) + .25y(.6x - .3x^2) =$

9. $250g(30g - 100h) + 440g(g + h) =$

10. $a^2b(bc^2 - abc^2) - a^3(4b + 3b^2c^2) =$

Name: _____ Date: _____

CHAPTER 2 REVIEW TEST

1. Which one of the following is a term?

    3xy   4y - 2y = 6y   8y² + 3x³

2. Which one of the following is an expression?

    4abc   4(x + 2) = 36   3ab + 4ab²

3. Which one of the following is an equation?

    3 + 5 = 8   4x - 3x   3xyz

4. If x = 7, then what is: 8x

5. If x = 6, then what is: $\frac{42}{x}$

6. Combine the like terms: $2xy^4 - 4x + 6xy^4 + x - 3y + xy^4 =$

7. Solve for m. $\sqrt{m} = 16$

8. $xy^3 \cdot x^2 \cdot y^4 =$

9. $4a \cdot 6a^2 \cdot b^3 \cdot b =$

10. Simplify the following: $3(2a + 4b) =$

11. Solve for x: $(x \cdot 4) + 5 = 17$

12. Solve for x: $3(9 + x) = 45$

Name: _____   Date: _____

Chapter 2 Review Test page 2

13. Write out the proper order of operations below:

_____ _____ _____ _____ _____ _____

Use the proper order of operations to solve the following 3 problems.

14. $[(9 \cdot 5) + 5] - 3^2 \cdot 6 + 4 =$

15. $(5^2 + 5 - 2^2) - \sqrt{81} \cdot 3 - 7 =$

16. $\dfrac{(4^2 + 3^2)}{x} = 5$

Use the Distributive Property of Multiplication to simplify the following.

17. $3rs(6r - 4s)$

18. $\dfrac{1}{2}\left(4a + \dfrac{5}{7}b\right)$

19. $4(6x + 3a) + 3(7a - 8x)$

20. $9a(7a + 4a) + a(a - b)$

21. $13a^2(ab + 4b) + 6a(2ab + b)$

22. $-11xy^2(9y - xy) - 5x^4(2y - x)$

CHAPTER 3
SLOPES
LESSON 20: GRAPHS

What you have learned, so far, may seem like worthless information, but there is a point to this math. One of the first ways you are going to use this knowledge is with *slopes*. A lot of people get lost in math once they get to slopes. It's not difficult, but it can be confusing, if not explained correctly.

First of all, let me tell you that slopes are not as complicated as they seem. Have you ever gone down a hill? Up a hill? Then you've been on a slope. Ever heard of a ski slope? It's a hill. A hill is a slope.

In higher math, graphs are used a lot. The lines that you draw on these graphs will either go uphill, downhill, or flat. That is the kind of slope we are talking about – the slope of the line drawn on a graph.

To understand slopes, you must first understand *graphs*. Have you ever used latitude and longitude lines on a globe? Those are the imaginary lines that go around the earth and up and down the earth, so you can find any point on the globe by finding where these lines intersect. That is how a GPS works. It finds where you are by using these imaginary lines. That is a type of *graph*, too.

Basically, a graph is a bunch of lines going up and down with a bunch of lines going side to side. Where these two lines intersect or cross each other, create a spot on the graph. Each spot can be labeled with *coordinates*. *Coordinates* look like this, (3, 2). They tell us where the two lines on the graph intersect. One number is for the up/down lines and the other number is for the left/right lines.

On the next page is a picture of another type of graph. This graph is plotting test scores. Look at the bottom of the graph to find each test. Go up to find the score for each test. It looks like Test 1 had a score of 10 and Test 3 had a score of 7. Can you tell what the score was for Test 2?

Test Scores

[Line graph showing: Test 1 = 10, Test 2 = 8, Test 3 = 7, Test 4 = 9]

That's right. Test 2 had a score of 8. This is how a basic graph works. You find one piece of information on the horizontal line and then you go up to get the other piece of information. There are many different types of graphs, but we are just going to learn about the kind of graph they use a lot in math an XY graph.

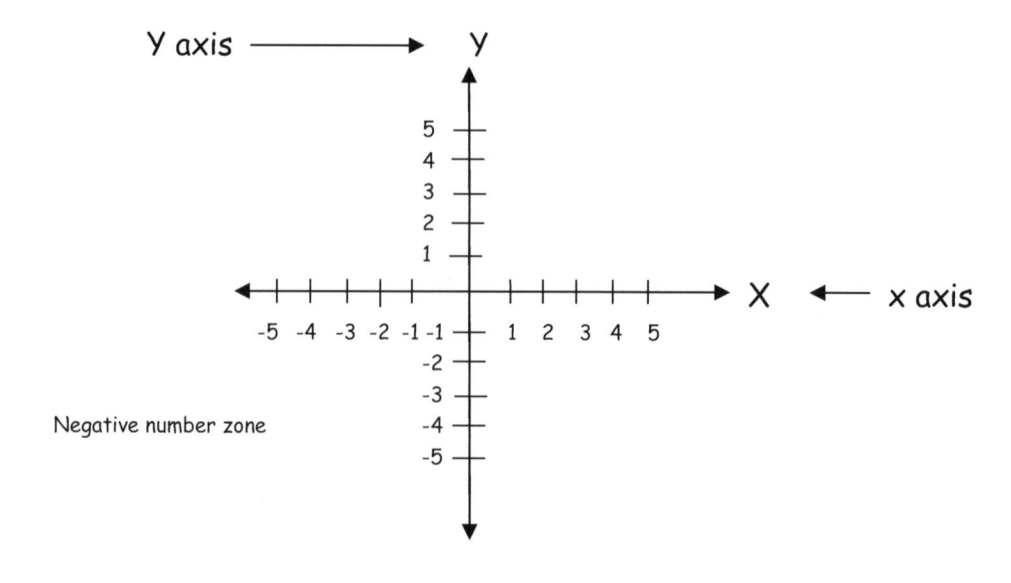

[XY axis diagram with Y axis labeled vertically from -5 to 5, X axis labeled horizontally from -5 to 5, with "Negative number zone" labeled on the lower left]

120

To find a point on the graph you look at a set of *coordinates*. I have made a list of 4 different sets of coordinate points. The first number is for the left/right line (x axis). And the second number is for the up/down line (y axis).

Look at the coordinates below.

$$(1, 1) \quad (2, 2) \quad (3, 3) \quad (4, 4)$$

I have plotted each of these points on the graph below and then I connected the dots with a line. Try to find the coordinate points above on the graph below. Find the first number on the x axis and then find the second number on the y axis. Follow those two lines until they meet. Find all 4 points on the graph.

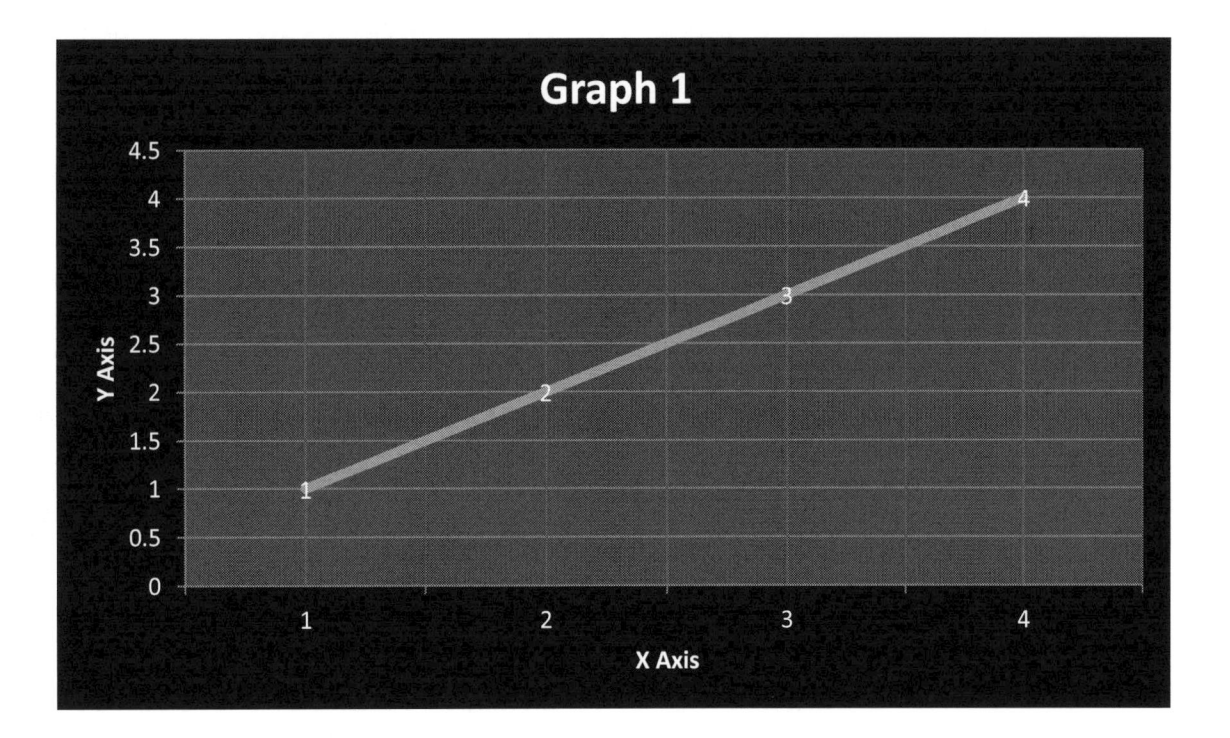

Graph 1

This particular line goes uphill, when it is drawn on a graph. It is not a real steep hill, but it certainly isn't a flat line either. To find out how steep of a line we have drawn, is to find the *slope* of the line. In this lesson you will learn how to calculate the slope of a line using a big mathematical formula.

But let me tell you a secret. It looks really complicated, but it's nothing more than two subtraction problems.

Before we start with the slope formula, let's make sure you completely understand graphs and coordinate points.

The graph on the last page had two number lines. The two number lines have names. The one that is going left to right is called the "x-axis." The number line that is going up and down is called the "y-axis." Together they create an "L" shape. The two axes (plural for axis) meet at the origin. The coordinate points for the origin are (0, 0).

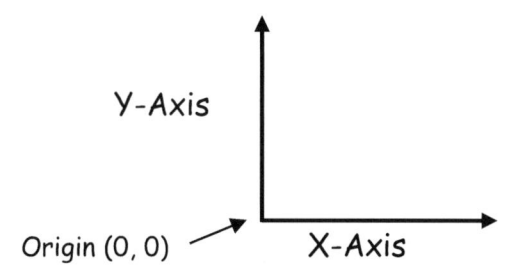

Y-Axis

Origin (0, 0)   X-Axis

Often you will need to find negative numbers on a graph. To show negative numbers we have to extend our axes to go beyond zero. Study the drawing below. It shows each number line going beyond zero or beyond the origin.

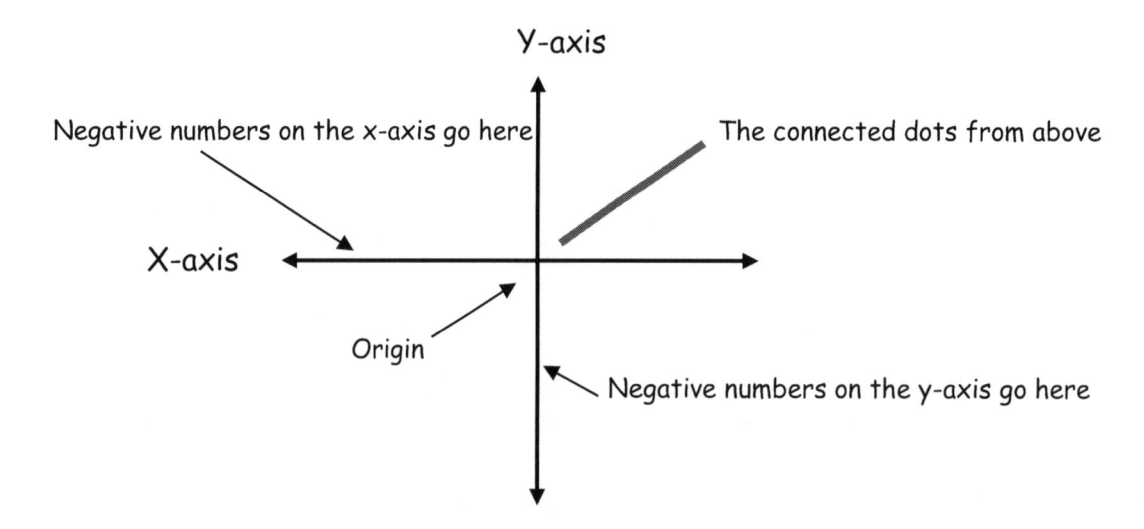

Y-axis

Negative numbers on the x-axis go here

The connected dots from above

X-axis

Origin

Negative numbers on the y-axis go here

That is a lot to remember, so let me break it down for you to make it easier to remember. When you read a sentence, you read from left to right. To read from right to left would be reading backwards, wouldn't it? Think of positive numbers on the x-axis as reading forward, to the right. And think of negative numbers as reading backwards.

On the y-axis, remember that going up, towards Heaven, is positive. Going down is negative. Negative numbers are less than zero, so they are below the origin.

Coordinates are like an address for each point on the graph. They tell you which two lines are intersecting to create the point. Sometimes you will see (x, y). That is just a way of writing a blank set of coordinates. It is up to us to fill in the x and y with numbers.

We will practice finding points on a graph using these two coordinates: (1, 2) and (3, 4).

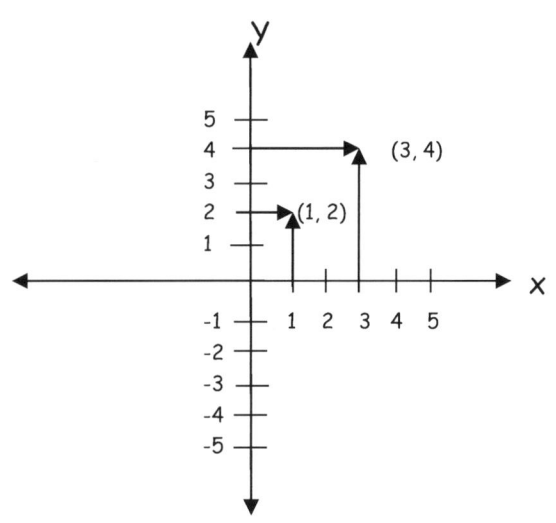

First, we will start with (1, 2). This means find the number 1 on the x-axis and then go up to number 2 on the y-axis. Draw a point where the 2 lines intersect and call it point (1, 2). Next, we will plot (3, 4) on the graph. Go over to 3 on the x-axis and then go up 4 spaces on the y-axis. Draw your point where the lines intersect. That's simple enough. Wouldn't you agree?

Sometimes when you are reading coordinates it is hard to remember which one is x and which one is y. Here is the way I remember. When I read a book, I **READ** from left to right, just like the **x**-axis. The **FIRST** number in a set of coordinates is for the x axis, so I created a phrase to help me remember which number is first. Use this phrase to help you remember: **"Read X first."**

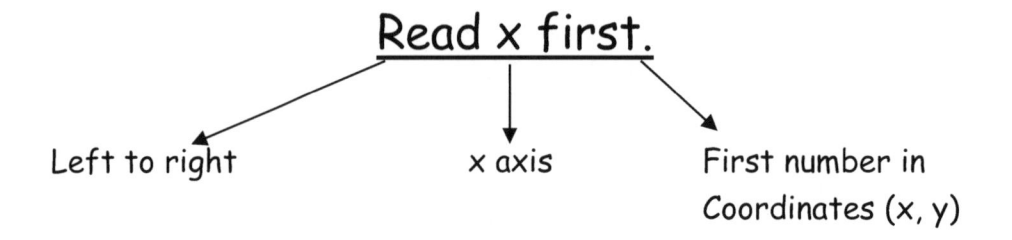

<u>Read x first.</u>

| Left to right | x axis | First number in Coordinates (x, y) |

Below is a graph with two points drawn on it. Can you name the coordinates for each point?

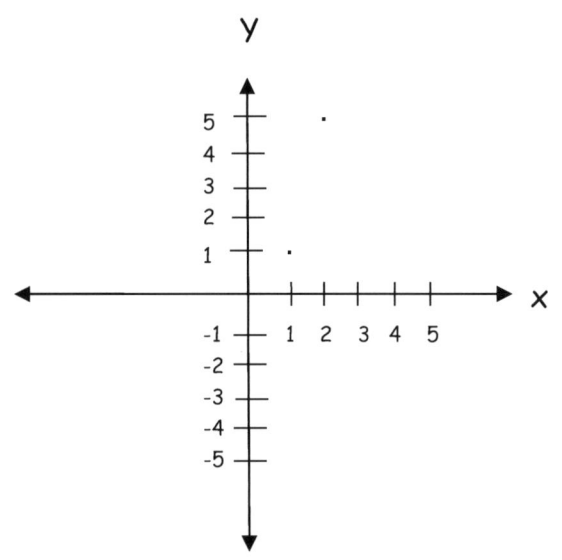

Did you get coordinate points (1, 1) and (2, 5)? If not, figure out where you went wrong. When we connect the dots, it draws a line. What this line looks like is a big deal to mathematicians. They want to know if this line is going uphill or downhill and how steep it is. That's what a slope is; how steep the line is on the graph. All kinds of information can be learned from the slope of a line. For now, you just need to know how to draw a line on a graph and how to figure out the slope of that line, which we will cover in the next lesson.

Name: _____   Date: _____

WORKSHEET 3-20

1. Plot the following coordinate points on the graph below:
   (6, 7) (4, 4) (2, 3) (-3, -2) (-2, -5)

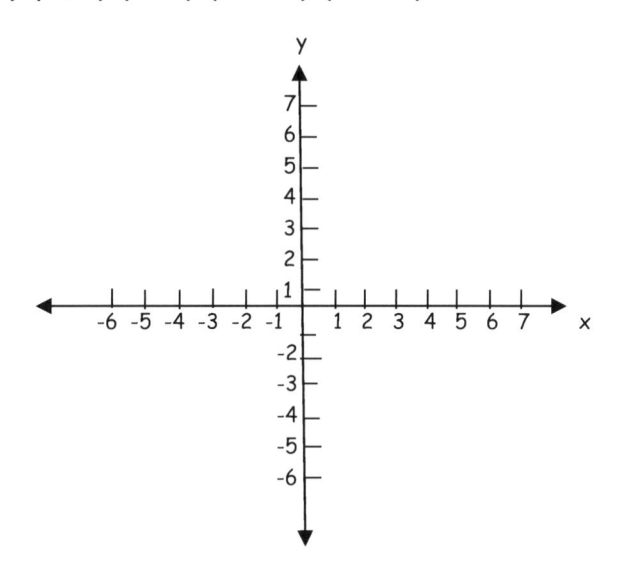

2. Below is a graph. Locate the 4 points and name them (x, y).

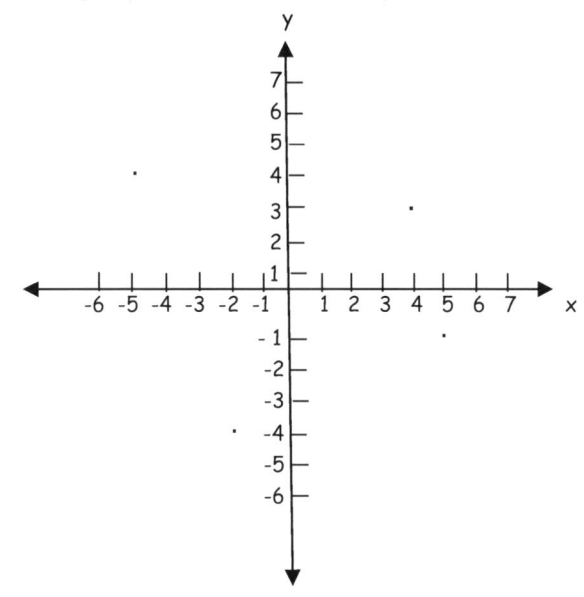

3. Which number in the coordinates (6, 7) refers to the y-axis. Does the y-axis go up and down or left to right?

WORKSHEET 3-20 page 2

4. Below is a graph. Plot the following points and then connect the dots to make a straight line. (-4, -4), (-2, -2), (0, 0), (1, 1) and (5, 5).

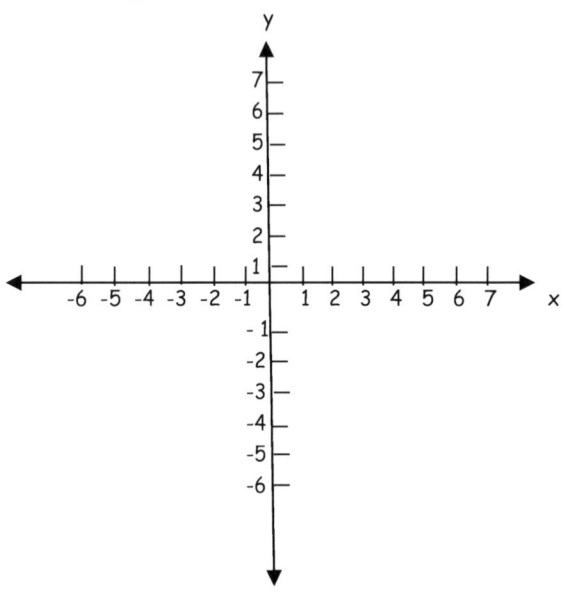

5. Look back at problem number 4. You drew a straight line to connect five different points. Name three other points that your line passes through.

6. Look at the graph above. If you have two negative coordinates such as (-3, -2) your point will always be located in:

   a. The upper right side.   c. The upper left side.
   b. The lower right side.   d. The lower left side.

7. What does the phrase "Read X First" mean?

LESSON 21: LINEAR EQUATIONS

Let's plot two sets of coordinates on a graph, so we can draw a line. We will draw in (1, 1) and (3, 3).

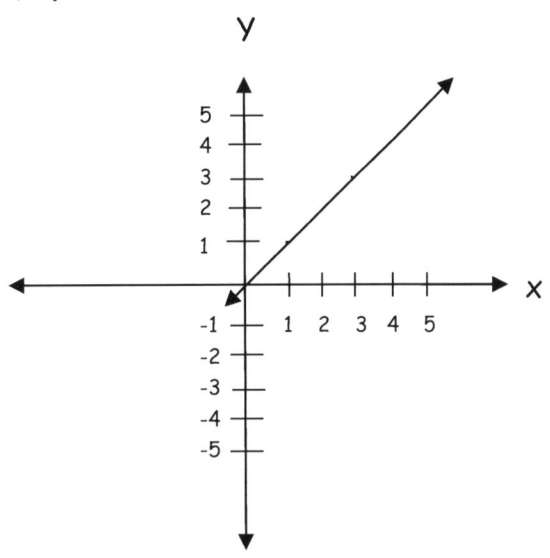

Now that we have plotted 2 points and drawn a line connecting them, we need to find out how steep of a line we have. Let me explain what I mean with a staircase.

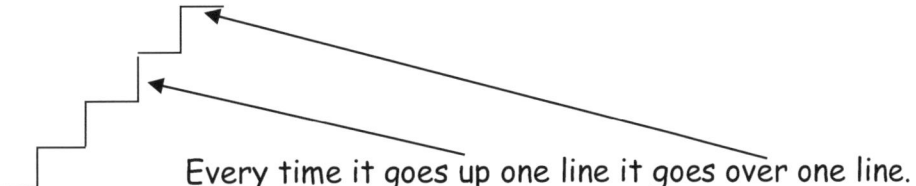

Every time it goes up one line it goes over one line.

If we drew a line on top of these stairs, they would have a slope of 1. That's because the line goes up 1, then over 1, then up 1, then over 1. That is a slope of 1. That is a way to look at it logically, but we need to learn how to figure the slope mathematically.

Since we already used the letters x and y in our graph, mathematicians decided to use the letter "m" to represent a slope. I don't know why they chose the letter "m" to represent a slope, but somebody, somewhere came up with "m," so that's what everyone uses for slopes. When we say m = 1, that means the staircase has a slope of 1. Don't get confused, so far this is all you've learned about slopes:

- A slope is a measurement of how steep a line is on a graph.
- You learned how to plot points on a graph using a set of coordinates.
- How to draw a line on a graph by connecting the dots.

If you understand those three points above, continue reading. Otherwise, go back to the beginning of this chapter.

Sometimes in math, we are given the coordinates to plot on a graph. But other times, we have to mathematically figure out the points. That's when your pre-algebra skills come in handy.

This time, instead of giving you coordinates to plot on a graph, I will just give you an equation and you have to figure out the coordinates yourself. Sounds kind of scary, so let me explain.

Do you remember when you learned how to "solve for x in terms of y?" When you "solve for x in terms of y," you end up with an answer like $x = y + 3$. Which means, give me a number for y and I'll give you the value of x.

Once we have numbers for x and y, we can plug them into (x, y) and BAM we've got a set of coordinates. If we get two or more of these coordinates, then we can draw a line on a graph.

We start by drawing a quick table, like the one on the next page. I put all of the x coordinates on one side and all the y coordinates on the other side. We don't know any of the coordinates yet. All we have is an equation, so we are going to make up some numbers for x and then we will use the equation to figure out y.

We are going to use the equation $y = x + 1$, to figure out several sets of coordinates. If you are a little confused, that's OK. Keep reading and it will get easier.

$$y = x + 1$$

Below is a table with some made up numbers for x. Next to that is an "equation machine."

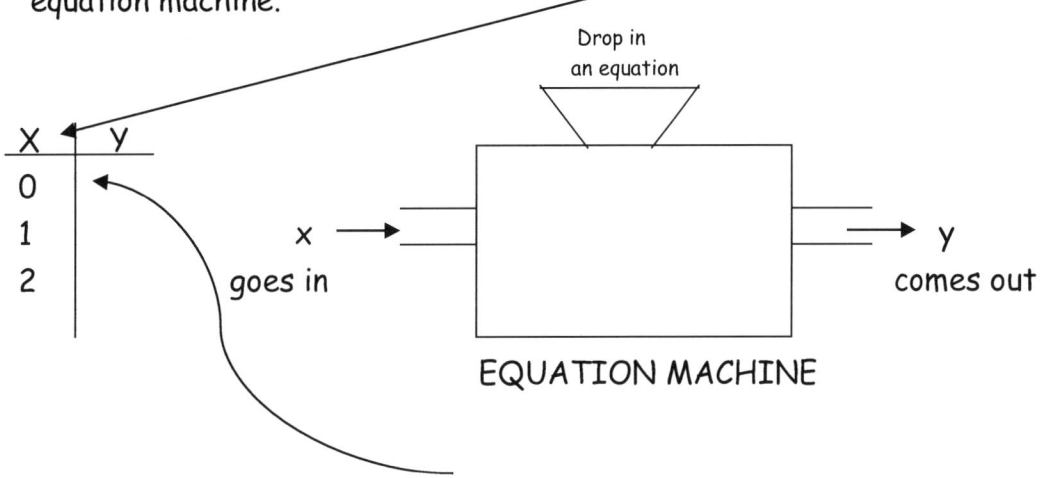

X	Y
0	
1	
2	

Drop in an equation

x goes in → EQUATION MACHINE → y comes out

Look at the table above. The first number I gave x is 0.
Put that x into the "equation machine" and let's see what we get for y.
Let's peek inside the machine. I'll explain each step.

Input   Equation   Math   Output
$x = 0$ → $y = x + 1$ → $y = 0 + 1$ → $y = 1$

I put an $x = 0$ in the machine. The 0 is plugged into the x in the equation. The machine does the math and gives us a value for $y = 1$. Let's add that to our table. The table below is saying that if your x coordinate is 0, then your y coordinate is going to be 1, according to the equation $y = x + 1$.

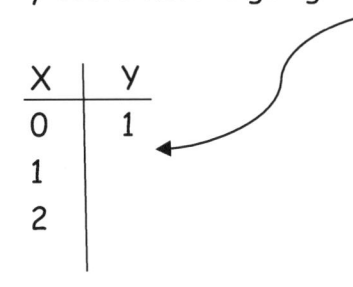

X	Y
0	1
1	
2	

Now let's find out what y will be if $x = 1$. Put $x = 1$ into the machine and see what we get for a y coordinate.

$x = 1 \longrightarrow y = x + 1 \longrightarrow y = 1 + 1 \longrightarrow y = 2$

Add that to our table.

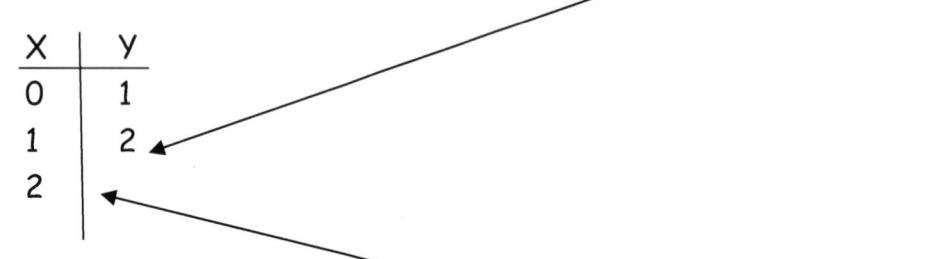

X	Y
0	1
1	2
2	

Put in the last x coordinate, 2. What do you get for y?

$x = 2 \longrightarrow y = x + 1 \longrightarrow y = 2 + 1 \longrightarrow y = 3$

Add y = 3 to our table and we'll have three sets of coordinates to plot on our graph.

X	Y
0	1
1	2
2	3

Our coordinates are (0, 1) (1, 2) and (2, 3).

Now we can plot these points on a graph and connect the dots.

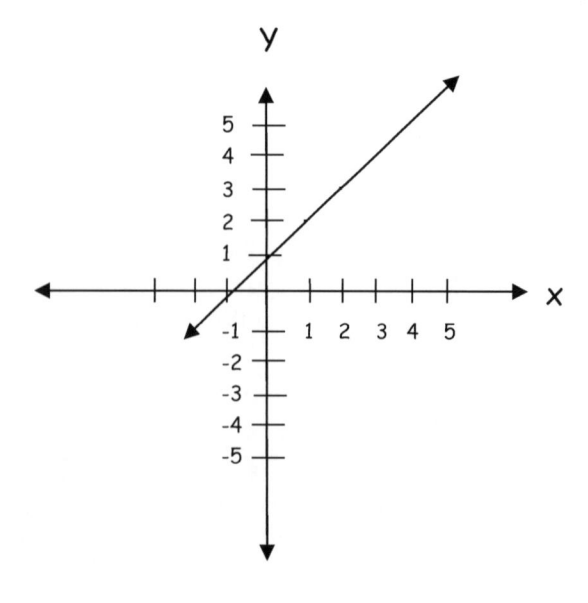

The graph on the last page shows the line we created with the *linear equation*, y = x + 1. It is a linear equation because that equation draws a line. Get it? A "line-ar" equation draws a "line." That last linear equation was easy. We barely even needed the equation machine to figure out y, so let's try a more difficult equation.

$$y = 2x - 1$$

Start by drawing a table full of some x values. We will use 0, 1, and 2 again. I could use any number for x, but if I put in 250, I'm going to need a pretty big graph to plot those points, so I'll stick with low numbers. Not only that, low numbers make the math easier.

x	y
0	
1	
2	

We need to put each of these numbers into our equation machine, so we can get some y coordinates. Starting with x = 0. Here is what goes on inside the machine.

x = 0 ⟶	y = 2x − 1	y = 2(0) −1	y = 0 − 1 ⟶	y = −1
Input x	Equation	Fill in x	Do the math	Output y

x = 1 ⟶	y = 2x − 1	y = 2(1) −1	y = 2 − 1 ⟶	y = 1
Input x	Equation	Fill in x	Do the math	Output y

x = 2 ⟶	y = 2x − 1	y = 2(2) −1	y = 4 − 1 ⟶	y = 3
Input x	Equation	Fill in x	Do the math	Output y

Fill in the table.

x	y
0	−1
1	1
2	3

Our coordinates are (0, −1), (1, 1) and (2, 3).

Plot those coordinates on the graph and then connect the dots.

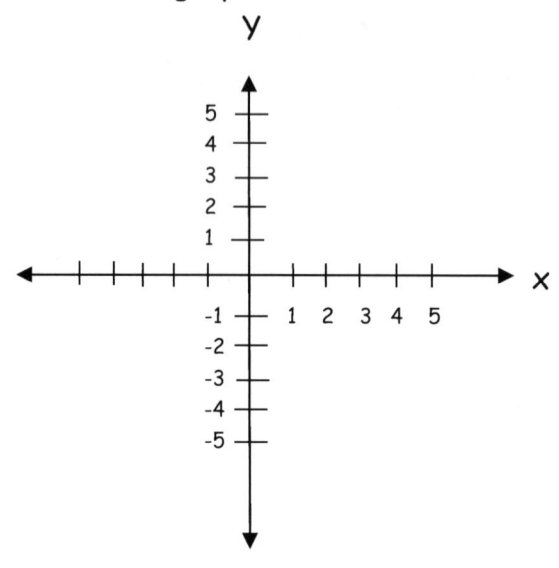

Name: _____   Date: _____

WORKSHEET 3-21

Use the following equations to find the missing y values for each table.

1. y = x + 3

x	y
1	
2	
3	

2. y = 3x + 1

x	y
1	
2	
3	

3. y = 2x -2

x	y
1	
2	
3	

4. y = 3x – 4

x	y
1	
2	
3	

5. y = 2x – 5

x	y
1	
2	
3	

LESSON 22: SLOPE OF A LINE

OK, you know what the slope of a line means. You know how to get some coordinates from an equation, and you can plot them on a graph. Now you need to learn how to *find* the slope of a line.

In this chapter you will learn 5 different "slope skills" for your math toolbox. Then we will loop them together, so you fully understand all the fundamentals of slopes. In algebra you will probably only work on one skill at a time, but we are going to connect all the skills together. It will make more sense that way.

To find the slope of a line you need to use a *formula*. A formula is like a math template. It has everything in the right order. It is just waiting for you to put in your numbers. If you wanted to find out how old someone was 5 years ago, you could use this formula.

$$x = a - 5$$

Fill in "a" with your age and x will equal your age 5 years ago. You could put anybody's age in this formula, and you would get their age 5 years ago. That's all a formula is; just letters waiting for your numbers.

This next formula is the formula everyone uses to figure out the slope of a line. It is a very important formula, and it is used a lot in math. It looks complicated, but it is actually very simple. A lot of people get very frustrated with math once they get to slopes because they assume it is complicated. It only looks and sounds complicated. It's actually really easy math, once you understand it, so don't over complicate it. Learn it and then teach it to someone else.

$$m = \frac{y_2 - y_1}{x_2 - x_1}$$

The formula above expects you to have two sets of coordinates to put in place of the x's and y's.

Coordinates are written as (x, y). When we have two sets of coordinates, we call the first set (x_1, y_1) and the second set (x_2, y_2). Don't let those little 1's and 2's confuse you. All that means is set 1 and set 2.

Below are 2 sets of coordinates (points on a graph). We will put them into the formula. This will give us the slope of the line that these two sets of coordinates would make, if we plotted them on a graph. Look below to find y_1. It is a 4. Follow the arrow to see where y_1 goes in the formula. That now becomes a 4. Fill in the rest of the letters with coordinates.

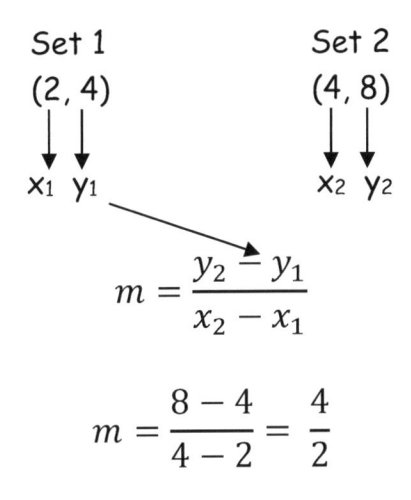

Set 1　　　　　Set 2
(2, 4)　　　　　(4, 8)
↓ ↓　　　　　　↓ ↓
x_1 y_1　　　　　x_2 y_2

$$m = \frac{y_2 - y_1}{x_2 - x_1}$$

$$m = \frac{8 - 4}{4 - 2} = \frac{4}{2}$$

Do the math. 8 - 4 = 4 and 4 - 2 = 2.

$m = \frac{4}{2}$　reduce that to　$m = 2$

The answer is 2. The slope of the line created when you plot (2, 4) and (4, 8) is 2. On the next page is a graph that shows the line created.

$$m = \frac{8-4}{4-2} = \frac{4}{2} = 2 \qquad\qquad m = \frac{4-8}{2-4} = \frac{-4}{-2} = 2$$

m = 2

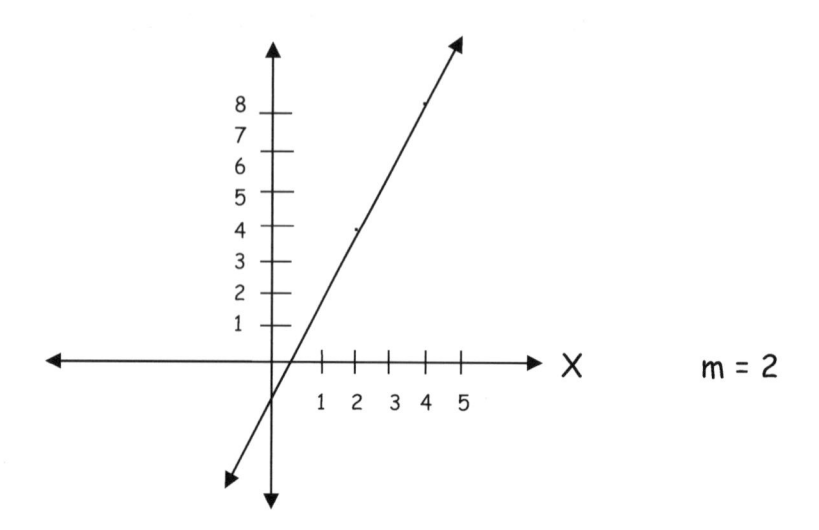

Had you accidentally subtracted that last problem backwards, here is the answer you would have got.

Formula filled in correctly (left) — Formula filled in backwards (right)

Surprisingly, if you fill out the formula backwards, you still get the right answer. Some teachers say you can subtract either way as long as you are consistent and keep the y's on top and the x's on the bottom. I prefer to stick to the formula.

Let's review everything we've learned about slopes, so far. Coordinates are two numbers grouped together with parentheses. The first number is the x-coordinate. The second number is the y-coordinate. We find these numbers on a number line called the x-axis and the y-axis. The first number is located on the x-axis and the second number is either up or down from there. Where these two lines meet, is where we draw our point.

Sometimes, we are given a linear equation instead of coordinates. We use this equation to find several sets of coordinates by using the "equation machine."

Once we plot all the points on the graph, we connect the dots. It will draw a straight line because it is linear. We can measure how steep the line is by using the slope formula. All we need are two sets of coordinates from your line to use the formula.

The slopes formula is $m = \dfrac{y_2 - y_1}{x_2 - x_1}$

Fill in the x and y's with the correct coordinates and do the math. You now have the slope of that line.

Sometimes your slope will be an improper fraction. For example, it is possible for your slope to equal $\dfrac{5}{4}$. That is an improper fraction. It can be converted into a mixed number, $1\dfrac{1}{4}$. When a line has a slope of $1\dfrac{1}{4}$, it means it is sloped just a little more than 1.

Other times you may end up with a negative slope. If either the numerator or denominator is a negative number, then the slope is negative. When a line has a negative slope, it means it's going downhill. Here is an example of a line with a negative slope. I did the math below to prove it.

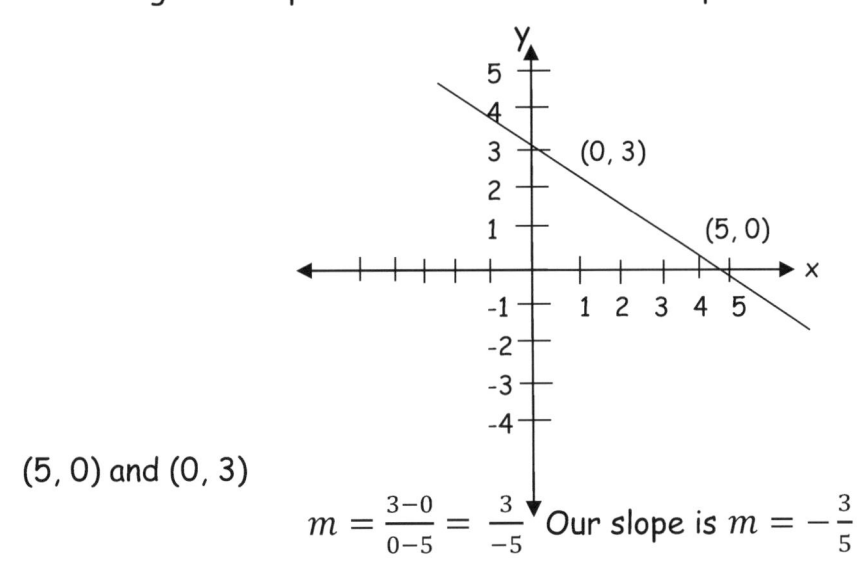

(5, 0) and (0, 3)

$$m = \dfrac{3-0}{0-5} = \dfrac{3}{-5} \quad \text{Our slope is } m = -\dfrac{3}{5}$$

A slope of negative $\dfrac{3}{5}$ means the slope of the line is less than 1 and it is going downhill.

That is an important fact to remember in algebra, so I'll say it again. If you are given two sets of coordinate points, you can tell if the line is going uphill or downhill by solving for the slope of the line. You don't even have to see the line to know. If your answer is a positive number, such as $m = 1$, then you know the line will go uphill.

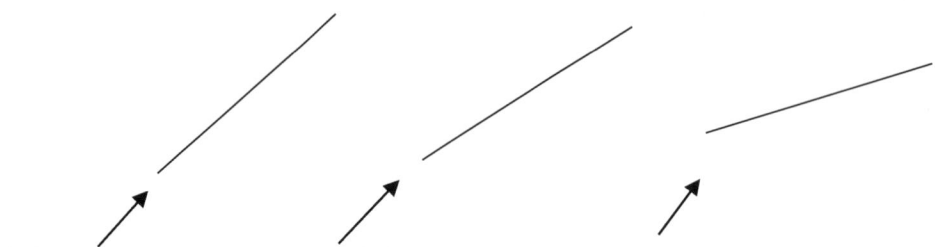

Like this……………….or this……………….or this. How steep the line is depends on the number. A slope such as $m = 2$ will be steeper than a line with a slope of 1, but remember, a positive slope will always go uphill.

On the other hand, a negative slope will always result in a downhill slope.

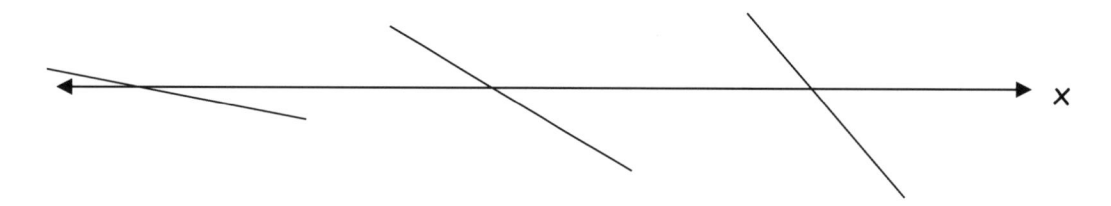

Think of yourself as walking along the x-axis. First, you step on the number 1. Your second step is on the number 2. Continue walking forward along the x-axis until you run into your line. Then ask yourself, "Does this line force me to walk downhill or uphill?"

Now I have a question for you. I've drawn a line on a graph, but you can't see it. When I solved for the slope of that line, my answer was $m = -\frac{1}{2}$. Which one of the three lines below is the only possible line that could have a slope of $-\frac{1}{2}$?

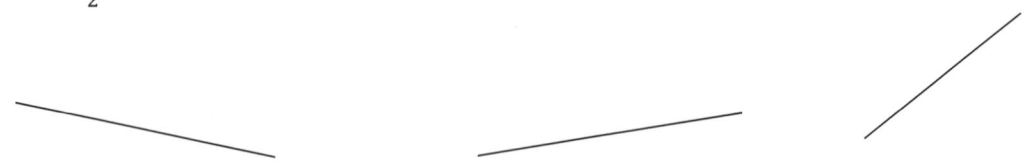

That's right! The first line is the only one that is going downhill. A negative slope will always result in a downhill slope.

Now let's talk about a line on a graph that is neither uphill nor downhill. For example, look at a straight horizontal line. Is it going uphill or downhill? Neither!

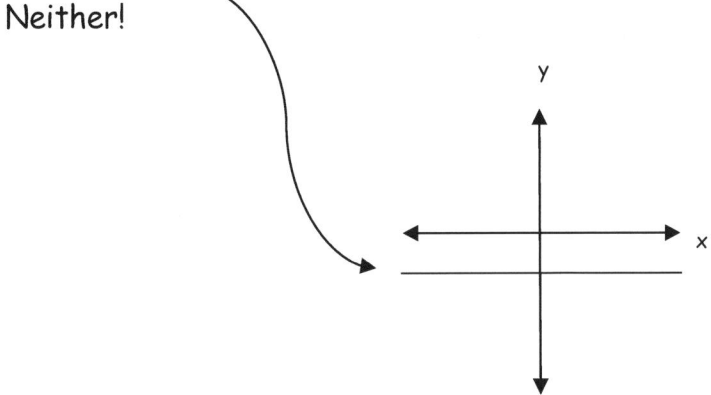

So then, what is the slope of that line? Let's do the math to find out. Two of the coordinate points on the flat line above are (-5, -3) and (4, -3). I'll plug those numbers into the slope formula.

$$m = \frac{y_2 - y_1}{x_2 - x_1}$$

$$m = \frac{-3 - -3}{4 - -5} = \frac{0}{9} \qquad m = \frac{0}{9}$$

Our slope equals $m = \frac{0}{9}$. What does that mean? Well, let's see, if you had zero of the nine pieces of pizza, then you would have zero pieces. The slope of a flat line is zero!

So far, you learned that a line on a graph will go either uphill, downhill, or it will be a flat horizontal line depending on the slope.

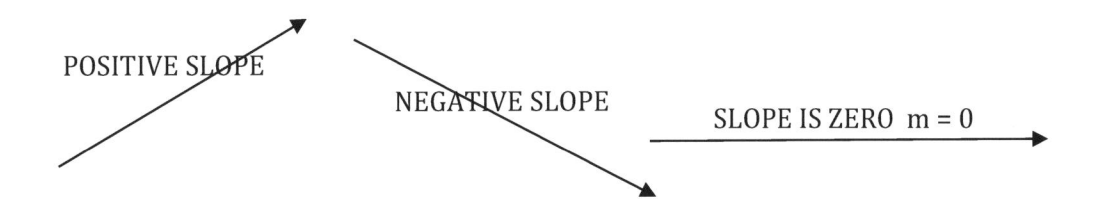

POSITIVE SLOPE   NEGATIVE SLOPE   SLOPE IS ZERO m = 0

There is just one more slope that we haven't talked about. That is the slope of a vertical line. What is the slope of the line below?

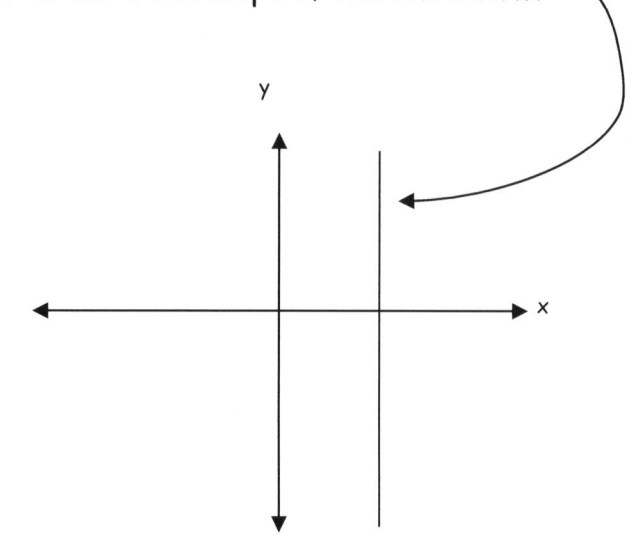

If you were walking along the x-axis, would this line force you to go uphill or downhill? Well...both, so what do we do? Let's do the math! Two of the coordinate points on this vertical line are (4, 4) and (4, -4).

$$m = \frac{-4-4}{4-4} = \frac{-8}{0} \qquad m = \frac{-8}{0}$$

What kind of number is $\frac{-8}{0}$? That doesn't even make sense! If it takes zero pieces to equal one and I have negative eight of those pieces...WHAT? Let's try to do the division. "Negative eight divided by zero." BUT WAIT!! You cannot divide by zero...it goes against all math logic because there is no number that will work. Whenever there is a zero in the denominator, it is "undefined" because the math is impossible!

$$\frac{3}{0} = undefined \qquad \frac{-5}{0} = undefined$$

If the slope of your line has a zero in the denominator, then you have a vertical line!

Practice finding the slope of a line by completing the next worksheet.

Name: _____   Date: _____

WORKSHEET 3-22

Find the slope of each line. The first one is done for you.

Slope formula is $m = \dfrac{y_2 - y_1}{x_2 - x_1}$

Coordinates	Fill in formula	Do the math
(2, 2) and (7, 4)	$m = \dfrac{4-2}{7-2}$	$m = \dfrac{2}{5}$

1. (-2, 4) and (4, 10)

2. (4, 1) and (10, 12)

3. (3, 2) and (5, 4)

4. (5, 7) and (3, -3)

5. (2, 5) and (7, 1)

6. (-3, 3) and (2, 3)

7. (-4, 2) and (3, -4)

WORKSHEET 3-22 page 2

The slope of ten different lines are listed below. State which direction the line would go in, if drawn on a graph (uphill, downhill, horizontal or vertical).

8. $m = \frac{8}{5}$ _____

9. $m = \frac{6}{12}$ _____

10. $m = -1$ _____

11. $m = 0$ _____

12. $m = \frac{0}{-5}$ _____

13. $m = 1\frac{1}{4}$ _____

14. $m = \frac{5}{0}$ _____

15. $m = \frac{1}{-2}$ _____

16. $m = 2$ _____

17. $m = \frac{9}{0}$ _____

LESSON 23: THE Y-INTERCEPT

There is one more formula to learn. Mathematicians are very curious to find out at which point your line crossed the y-axis. This is called the *y-intercept*. Look at the graph below. The line intercepts the y-axis at positive 2. Try not to be confused because it is really just that simple. Mathematicians find it important to know where this line crossed the y-axis. And they decided to call that point "b."

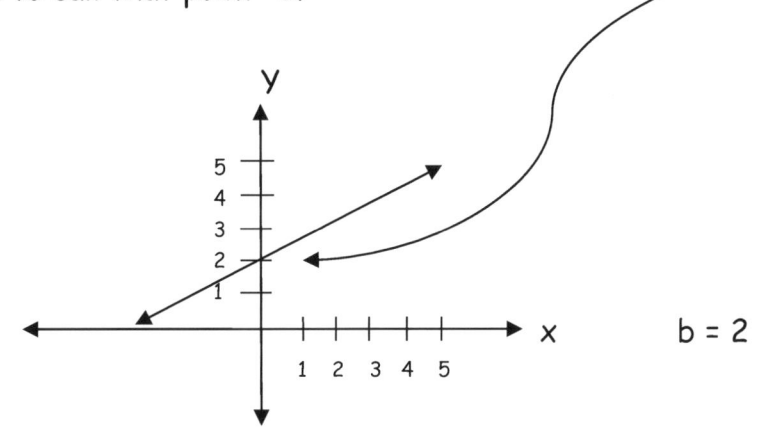

$$b = 2$$

The formula we use to find out where our line passes through the y-axis is in my toolbox, somewhere. Here it is.

$$y = mx + b$$

This formula is a big deal in the math world. You MUST learn this formula inside and out before you can succeed in algebra, so let's learn it now.

First, we will pull it apart and identify each letter. You already know three of them.

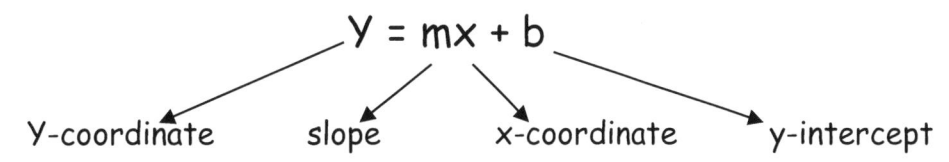

$$y = mx + b$$

Y-coordinate   slope   x-coordinate   y-intercept

Now let's find some numbers to put in place of y, m, and x, so we can figure out b. We will use the coordinates (2, 1) and (3, 2). Do you know what the slope is for the line created using these coordinates? Me either. We're going to have to figure that out first, so we can fill in the "m" in our formula.

Let's do the math. We will need our slope formula to figure that out. So, let me dig through my math toolbox and get it. Here it is.

Slope formula is... $m = \dfrac{y_2 - y_1}{x_2 - x_1}$

The coordinates are (2, 1) and (3, 2).

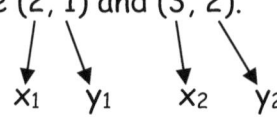

x_1 y_1 x_2 y_2

Fill in the formula $m = \dfrac{2-1}{3-2}$ Do the math. $m = \dfrac{1}{1}$ Reduce. $m = 1$

Now that we know our slope, let's go back to the formula y = mx + b and fill in all the numbers. We can pick either one of the two sets of coordinates. It doesn't matter, you will get the same answer either way. We will use (2, 1) for x and y, and m = 1.

(2, 1)   m = 1

y = mx + b

1 = 1 · 2 + b

Simplify the equation.   1 = 2 + b
Get b by itself.   -2 - 2
                          -1 = b

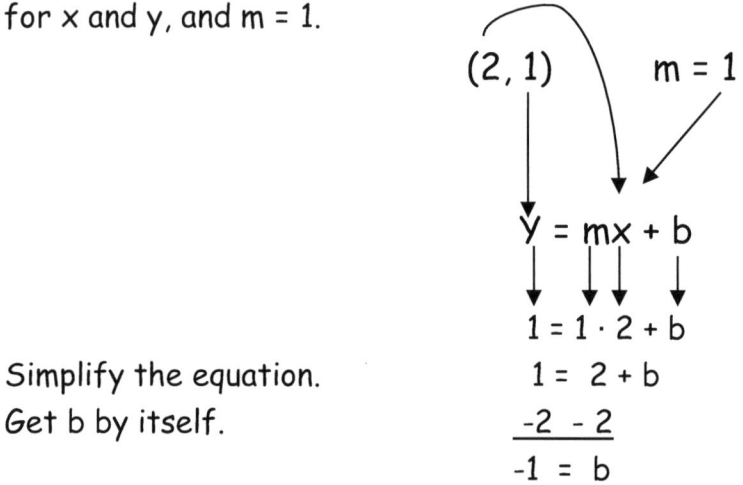

There we go. We figured out b. Remember b is the point on the y-axis where our line crosses through. Mathematicians know that our line HAS TO pass through the y-axis at some point and they want to know where. Was it at the 5 or -4? They want to know, so we just figured it out for them. It crosses the y-axis at -1. That should make them happy.

I plotted those coordinates on the next graph and then drew connected the dots with a line to see if it really does intercept the y-axis at -1. Remember, on the y-axis negative numbers are below the center.

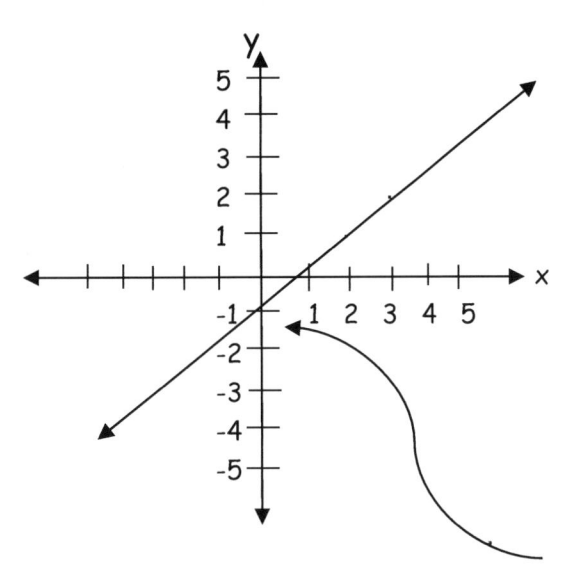

How about that? It really does intercept the y-axis at -1. Let's try it again using the other set of coordinates. Last time we used (2, 1) to solve for b. This time we will use the other point (3, 2). We should get the same answer because it is the same line with the same slope.

First, we need to find the slope of the line. We already have the coordinates. They are (3, 2) and (2, 1) we will need to use the slope formula to find our slope.

Slope formula is $m = \frac{y_2 - y_1}{x_2 - x_1}$

Our coordinates are (3, 2) and (2, 1)

x_1 y_1 x_2 y_2

Fill in the formula with those numbers.

$m = \frac{1-2}{2-3}$

$m = \frac{-1}{-1}$ -1 ÷ -1 = 1   so $m = 1$, just like last time.

OK, now we have our slope, $m = 1$, and we have our x and y (3, 2). Let's figure out b or, in other words, the y-intercept. We use y = mx + b to find b.

$$y = mx + b$$

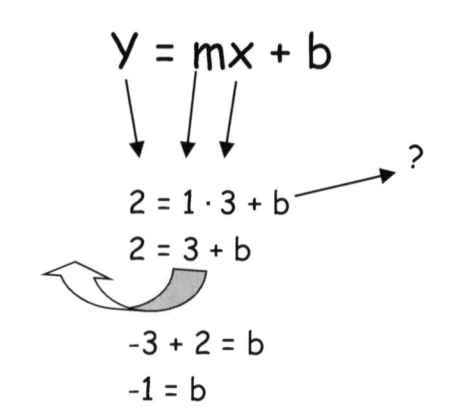

Do the math.

$2 = 1 \cdot 3 + b$ ⟶ ?
$2 = 3 + b$

$-3 + 2 = b$
$-1 = b$

Look at that, we got the same answer. The y-intercept or "b" is still -1.

That should have been really simple. If it was difficult for you, read this lesson again and don't feel bad; most people have a hard time learning slopes.

If it was easy for you, complete the next worksheet.

Name: _____   Date: _____

WORKSHEET 3-23

Below is a set of coordinates and a slope. Fill in the formula with the numbers given. Then simplify the equation you wrote and solve for b. The first one is done for you.

$$y = mx + b$$

(x, y)	Slope	Formula filled in	Simplified	Solve for b
(2, 4)	$m = \frac{1}{2}$	$4 = \frac{1}{2} \cdot 2 + b$	$4 = 1 + b$	$3 = b$
(0, 3)	$m = \frac{1}{2}$			
(1, 2)	$m = 1$			
(2, 3)	$m = 1$			
(5, 0)	$m = 4$			
(-2, 3)	$m = 0$			
(-4, -3)	$m = \frac{2}{3}$			
(5, -2)	$m = 2$			

LESSON 24: CREATING A LINEAR EQUATION

So far, learning slopes is pretty simple math, but let's review. You start with an equation like $y = x + 1$. From there, you draw a table and get some numbers for x and y. You plot those numbers on a graph and draw a line. Next, you figure out the slope of that line by using the slope formula. From there, you can solve for b, also known as, the y-intercept.

Good news, there is only one more thing to learn about slopes. Earlier you were given an equation. From there you came up with coordinates, then a slope, and ultimately a y-intercept. This time you will be given a set of coordinates and YOU have to come up with the equation. Let's give it a try.

Coordinates given:

(3, 1) and (7, 6)

First, we need to find the slope of this line. We use the slope formula for that.

$$m = \frac{y_2 - y_1}{x_2 - x_1}$$

$$m = \frac{6-1}{7-3} = \frac{5}{4}$$

The slope is $\frac{5}{4}$ and the coordinates are (3, 1) and (7, 6). All we need now is the y-intercept. We will use the y-intercept formula to find it.

$$y = mx + b$$

$$1 = \frac{5}{4} \cdot 3 + b$$

Let's do the math.

$$1 = \frac{5}{4} \cdot 3 + b$$

Turn 3 into a fraction to multiply.

$$1 = \frac{5}{4} \cdot \frac{3}{1} + b$$

Multiply the fractions.

$$1 = \frac{15}{4} + b$$

Swing it over to the other side and change the sign.

$$1 - \frac{15}{4} = b$$

Do the math.

$$\frac{4}{4} - \frac{15}{4} = -\frac{11}{4} = b$$

OK, now we have all the numbers that we need to create an equation.

$$m = \frac{5}{4} \qquad\qquad b = -\frac{11}{4}$$

The reason we are creating this equation is so you can find any point on the line. Let's put our numbers into the y-intercept formula, $y = mx + b$. We don't fill in the "y" or the "x" because we are looking for a blank equation that we can put into the equation machine. Once I fill in the m and the b, our equation will look like this:

$$y = \frac{5}{4}x - \frac{11}{4}$$

Since $\frac{11}{4}$ is a negative number, we subtract b, instead of add b.

With this equation we can determine as many points on our line as we'd like. Just make a table, come up with an x-coordinate and then use this equation to get your y-coordinate.

Let's try another one. This time I will give you one set of coordinates and the slope of a line. From there, all you need to find is b...and you've got all the numbers you need to form a linear equation.

$$\text{Coordinates} \qquad \text{Slope}$$
$$(3, 4) \qquad\qquad m = 2$$

We need to figure out b, the y-intercept. Let's fill in y = mx + b with the numbers we have above, so we can solve for b.

$$y = mx + b$$
$$4 = 2 \cdot 3 + b$$

$$4 = 6 + b$$

$$4 - 6 = b$$

$$-2 = b$$

Now we have our y intercept and our slope. With this information we can create an equation for our line. Fill in m and b with our numbers.

$$y = mx + b$$
$$y = 2x - 2$$

Since our y-intercept is a negative number, we subtract b, instead of add b. Remember, we don't fill in the y and the x, just m and b.

Name: _____ Date: _____

WORKSHEET 3-24

1. I will give you the slope and the y-intercept for several different lines. You practice filling in the formula, "y = mx + b" with these numbers, so we have an equation for each line. The first one is done for you.

$$y = mx + b$$

Slope	Y intercept	Equation
m = 2	b = 3	y = 2x + 3
m = 3	b = 1	
m = 6	b = 0	
m = $\frac{1}{2}$	b = 4	
m = 1	b = -3	
m = 5	b = -1	

2. Fill in the blanks. Write what each letter in the y-intercept formula represents.

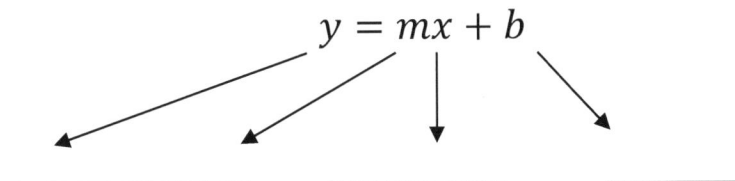

$$y = mx + b$$

_____   _____   _____   _____

3. What is the y-intercept?

151

LESSON 25: SLOPES REVIEW

Let's review everything we've learned about slopes, one more time. Once you understand all the elements of learning slopes, you will realize that the math isn't the difficult part- the concept is the challenge.

Basically, there are 5 different skills you need to learn to conquer slopes:
1. Learning to find coordinates from an equation like y = x + 1.
2. Plotting the points on the graph and connecting the dots.
3. Finding the slope of the line using the slope formula.
4. Finding the y-intercept.
5. Creating an equation for a line.

The first skill is finding coordinates from an equation. We did that by drawing up a table and making up some numbers for x. Then we used the equation to find some y-coordinates. If you can do that, one down, four to go.

The second skill is plotting points on a graph. Remember our phrase "Read x First." This means the **first** number in a set of coordinates is for the **x-axis**. So, if you see a set of numbers like (3, 4), know that 3 is the x-coordinate. The x-axis is the line that goes left to right. The same direction you **read**.

To find a point on a graph, start with the x-axis. Look at the first number in the set of coordinates and then find that number on the x-axis. Look at the second number in the coordinates. If it is a positive number, go up from your x position to that second number on the y-axis. If the second number is negative, as in (3, -5), then you will go down from your x position. Remember this, going up to Heaven is positive, going down is negative.

The third skill is finding the slope of a line. In order to find the slope of a line on a graph, we must have two sets of coordinates. From there, just fill in the slope formula with your coordinates. The first set is marked with little 1's and the second set is marked with little 2's. Fill in the slope formula with your coordinates and subtract.

$$m = \frac{y_2 - y_1}{x_2 - x_1}$$

The fourth skill is finding the y-intercept. If you know the slope of a line and at least one set of coordinates, you can figure out where that line intercepts the y-axis. We use this formula to figure out b, the y-intercept.

$$y = mx + b$$

Fill in the y, x, and m with your numbers and solve for b to get the y-intercept.

Finally, the last skill you need to master is creating an equation for a line. You are given two sets of coordinates and from there you can figure out the equation of the line. With this equation you can go back to skill number one and start over. We just completed the loop.

If all of this information makes sense to you, you are ready for the test. If you are even a little bit confused, just read this lesson again. You'll get it eventually. It's not hard math; it's just weird math.

Name: _____   Date: _____

CHAPTER 3 REVIEW TEST

Complete the 5 skills used to solve slopes.

1. Below is an equation. Use this equation to come up with 3 sets of coordinates for a line. Fill in the table with the y-coordinates.

    $$Y = 2x - 1$$

x	y
2	
3	
4	

2. Draw a graph and plot the points you found from the first problem. Connect the dots.

3. Use the slopes formula to find the slope of the line you drew on your graph.

    $$m = \frac{y_2 - y_1}{x_2 - x_1}$$

4. Use the formula below to find the y-intercept of your line. If your graph is drawn perfectly, you will see where the line intercepts the y-axis.

    $$Y = mx + b$$

5. Make an equation for a line that has these two coordinates: (2, 4) and (4, 8)

If you did well on that test: congratulations! You are now among the few people who can say they actually understand slopes.

Name: _____   Date: _____

FINAL TEST

Solve for x.

1. $\frac{1}{2}x \cdot 8 = 32$
2. $x + 7^2 = 62$
3. $5^3 = x$

4. $x(5 + 8) = 78$
5. $5 + 6 \times 9 - x = 51$

6. $\frac{x}{9} = \frac{4}{6}$
7. $\frac{1}{3}x = \frac{7}{24}$
8. $\sqrt{x} = 8$

9. $\frac{75}{x} = 15$
10. $x^2 = 81$
11. $\frac{x}{7} = 10$

Use the Distributive Property of Multiplication to simplify the following.

12. $3x(2x + 4y) + 7y(x - y) =$

13. $-x(7x - 3y) + 6x(-y + 5x) =$

14. $-3(a + b) + a(-2 + b) =$

15. Draw 3 tables and find 3 coordinates from these linear equations:

    $y = x - 4$ $y = 2x + 1$ $y = 3x - 1$

16. Draw a graph and plot 3 points using this linear equation: $y = 2x - 1$.

17. Look at the graph you drew for problem 16. Connect the dots and find the slope of that line by using the slopes formula.

$$m = \frac{y_2 - y_1}{x_2 - x_1}$$

Name: _____ Date: _____

Final Test page 2

18. Find the y-intercept of the line from problem 17 by using the y-intercept formula:

$$y = mx + b$$

Look at the coordinates below and create a linear equation.

19. $(2, 3)\ (3, 4)$

20. $(2, 1)\ (5, 5)$

21. $(2, 3)\ (4, 1)$

22. What does it mean when the slope of a line is a negative number?

23. What does a linear equation make?

24. What is the y-intercept?

25. What does each letter in the formula below represent?

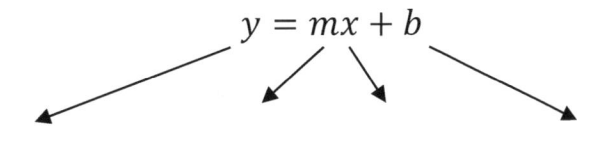

$$y = mx + b$$

_____   _____   _____   _____

ANSWERS

WORKSHEET 3-1

1. $4 + x = 24$
   $x = 24 - 4$
   $x = 20$

2. $x + 14 = 21$
   $x = 21 - 14$
   $x = 7$

3. $2 + x = 12$
   $x = 12 - 2$
   $x = 10$

4. $x + 72 = 172$
   $x = 172 - 72$
   $x = 100$

5. $x - 33 = 54$
   $x = 54 + 33$
   $x = 87$

6. $9 + x = 62$
   $x = 62 - 9$
   $x = 53$

7. $4 + x = 0$
   $x = 0 - 4$
   $x = -4$

8. $x + 25 = 100$
   $x = 100 - 25$
   $x = 75$

9. $x - 8 = 176$
   $x = 176 + 8$
   $x = 184$

10. $10 + x = 310$
    $x = 310 - 10$
    $x = 300$

11. $x + 38 = 44$
    $x = 44 - 38$
    $x = 6$

12. $x - 8 = 34$
    $x = 34 + 8$
    $x = 42$

13. $x - 8 = 5$
    $x = 8 + 5$
    $x = 13$

14. $x - 14 = -2$
    $x = -2 + 14$
    $x = 12$

15. $9 + x = -1$
    $x = -1 - 9$
    $x = -10$

16. $11 + x = -7$
    $x = -7 - 11$
    $x = -18$

17. $4 + x = 2$
    $x = 2 - 4$
    $x = -2$

18. $x - 3 = 27$
    $x = 27 + 3$
    $x = 30$

19. $x + 36 = 36$
    $x = 36 - 36$
    $x = 0$

20. $4 + x = -26$
    $x = -26 - 4$
    $x = -30$

21. $-33 = 7 + x$
    $-33 - 7 = x$
    $-40 = x$

Solve for x.

22. $43 + x = 57$
    $x = 57 - 43$
    $x = 14$
    $x = 14$ baseball cards

23. $9 + x = 17$
    $x = 17 - 9$
    $x = 8$ gallons

ANSWERS: WORKSHEET 3 – 1a

1. $-3 + x = -12$
   $x = -12 + 3$
   $x = -9$

2. $x - 14 = -28$
   $x = -28 + 14$
   $x = -14$

3. $12 + x = 4$
   $x = 4 - 12$
   $x = -8$

4. $10 + x = 5$
   $x = 5 - 10$
   $x = -5$

5. $x + 8 = -48$
   $x = -48 - 8$
   $x = -56$

6. $62 = x + 9$
   $x = 62 - 9$
   $x = 53$

7. $-44 = x - 18$
   $-44 + 18 = x$
   $-26 = x$

8. $x + 25 = -100$
   $x = -100 - 25$
   $x = -125$

9. $x - 10 = -17$
   $x = -17 + 10$
   $x = -7$

10. $x + 5 = -55$
    $x = -55 - 5$
    $x = -60$

11. $x + 7 = -12$
    $x = -12 - 7$
    $x = -19$

12. $8 + x = -20$
    $x = -20 - 8$
    $x = -28$.

13. $x - 27 = -15$
    $x = -15 + 27$
    $x = 12$

14. $-\frac{1}{2} + x = -.5$
    $x = -.5 + \frac{1}{2}$
    $x = 0$

15. $-.75 + x = .25$
    $x = .25 + .75$
    $x = 1$

16. $x + .3 = -5.3$
    $x = -5.3 - .3$
    $x = -5.6$

17. $\frac{1}{2} + x = 5$
    $x = 5 - \frac{1}{2}$
    $x = 4\frac{1}{2}$

18. $x - \frac{1}{4} = -6\frac{3}{4}$
    $x = -6\frac{3}{4} + \frac{1}{4}$
    $x = -6\frac{1}{2}$

19. Sarah is trying to break the record for doing the most one-handed cartwheels on a balance beam without falling. Right now, the record is 71, so she needs to get to 72 to break the record. She has done 15 cartwheels, so far. How many more does she need to do to break the record? Use algebra to solve for x in the equation below.

    $15 + x = 72$ $x = 72 - 15$ $x = 57$ cartwheels

20. Robin needs to keep track of the water level at Lake Welch. At the end of the summer, the water level was low. It measured 30 inches below the desired level. After a week of rain, the water level rose and is now only 16 inches below the desired level. How many inches did it rain?

    $-30 + x = -16$ $x = -16 + 30$ $x = 14$ inches of rain

ANSWERS: WORKSHEET 3-2

Solve for x.

1. $4x = 16$
   $x = 4$
2. $5x = 15$
   $x = 3$
3. $3x = 21$
   $x = 7$
4. $9x = 54$
   $x = 6$

5. $8x = 56$
   $x = 7$
6. $2x = 12$
   $x = 6$
7. $8x = 72$
   $x = 9$
8. $3x = 12$
   $x = 4$

9. $9x = 63$
   $x = 7$
10. $4x = 28$
    $x = 7$
11. $5x = 105$
    $x = 21$
12. $11x = 176$
    $x = 16$

13. $10x = 210$
    $x = 21$
14. $12x = 144$
    $x = 12$
15. $8x = 24$
    $x = 3$
16. $6x = 24$
    $x = 4$

17. $1/2 x = 4$
    $x = 8$
18. $3x = 42$
    $x = 14$
19. $9x = 81$
    $x = 9$
20. $11x = 154$
    $x = 14$

21. $4x = 32$
    $x = 8$
22. $7x = 42$
    $x = 6$
23. $6x = 48$
    $x = 8$
24. $6x = 36$
    $x = 6$

25. $2x = 100$
    $x = 50$
26. $4x = 36$
    $x = 9$
27. $9x = 18$
    $x = 2$
28. $49 = 7x$
    $7 = x$

29. $9 = 3x$
    $3 = x$
30. $27 = 3x$
    $9 = x$
31. $30 = 6x$
    $5 = x$
32. $48 = 12x$
    $4 = x$

33. $11x = 33$
    $x = 3$
34. $12x = 60$
    $x = 5$
35. $14x = 56$
    $x = 4$
36. $100x = -500$
    $x = -5$

ANSWERS: WORKSHEET 3-2a

1. $6x = 3$
$x = \frac{3}{6}$
$x = \frac{1}{2}$

2. $\frac{1}{2}x = 8$
$x = 8 \div \frac{1}{2}$
$x = 16$

3. $5x = 3.75$
$x = 3.75 \div 5$
$5\overline{)3.75}$
$x = .75$

4. $-5x = 25$
$x = 25 \div -5$
$x = -5$

5. $-\frac{1}{8}x = 1$
$x = 1 \div -\frac{1}{8}$
$x = \frac{1}{1} \times -\frac{8}{1}$
$x = -8$

6. $-2x = -34$
$x = -34 \div -2$
$x = 17$

7. $-5 + x = 2\frac{1}{2}$
$x = 2\frac{1}{2} + 5$
$x = 7\frac{1}{2}$

8. $10x = -6\frac{1}{4}$
$x = -6\frac{1}{4} \div \frac{10}{1}$
$x = -\frac{25}{4} \times \frac{1}{10}$
$x = -\frac{25}{40}$ $x = -\frac{5}{8}$

9. $8x = 64$
$x = 8$

10. $-7x = -49$
$x = -49 \div -7$
$x = 7$

11. $16x = -4$
$x = -4 \div 16$
$x = -\frac{1}{4}$

12. $72 = -9x$
$72 \div -9 = x$
$-8 = x$

13. $-3\frac{1}{8} + x = 2$
$x = 2 + 3\frac{1}{8}$
$x = 5\frac{1}{8}$

14. $8\frac{3}{8}x = -8\frac{3}{8}$
$x = -8\frac{3}{8} \div 8\frac{3}{8}$
$x = -1$

15. $12 + x = 4$
$x = 4 - 12$
$x = -8$

16. $x - \frac{5}{6} = \frac{2}{12}$
$x = \frac{2}{12} + \frac{5}{6}$
$x = 1$

17. $\frac{1}{2}x = -10$
$x = -20$

18. $8 = \frac{1}{2}x$
$x = 16$

19. $14x = 7$
$x = \frac{1}{2}$

20. $-2x = 4$
$x = -2$

21. A 5-inch hamburger patty shrinks down to $\frac{3}{4}$ that size when cooked. Here is the math, $\frac{3}{4} \cdot 5 = x$ $\frac{15}{4} = x$. But Collin wants the cooked burger to be exactly 3". What size hamburger patties should he make? Use algebra to figure it out.

$$\frac{3}{4}x = 3 \text{ inches}$$

$x = 3 \div \frac{3}{4}$ $x = \frac{3}{1} \times \frac{4}{3}$ $x = 4 \text{ inches}$

ANSWERS: WORKSHEET 3-3a

1. $\frac{x}{5} = 8$
   $x = 40$

2. $\frac{x}{9} = 7$
   $x = 63$

3. $\frac{x}{10} = \frac{1}{2}$
   $x = \frac{1}{2} \times \frac{10}{1}$ $x = 5$

4. $\frac{x}{8} = 2\frac{3}{4}$
   $x = \frac{11}{4} \times \frac{8}{1}$ $x = \frac{22}{1}$ $x = 22$

5. $\frac{x}{-8} = \frac{5}{8}$
   $x = \frac{5}{8} \cdot -\frac{8}{1}$ $x = -5$

6. $\frac{7}{16} = \frac{x}{24}$
   $\frac{7}{16} \cdot \frac{24}{1} = x$ $x = \frac{21}{2}$ $x = 10\frac{1}{2}$

7. $-9 = \frac{x}{3}$
   $x = -27$

8. $\frac{x}{4} = -3\frac{1}{2}$
   $x = -\frac{7}{2} \cdot \frac{4}{1}$ $x = -\frac{28}{2}$ $x = -14$

9. $18 = \frac{x}{\frac{1}{2}}$
   $x = \frac{18}{1} \cdot \frac{1}{2}$ $x = 9$

10. $\frac{x}{3} = -4\frac{7}{8}$
    $x = -\frac{39}{8} \times \frac{3}{1}$ $x = -\frac{117}{8}$ $x = -14\frac{5}{8}$

11. Connor is playing in a championship football game. His team has a score of 56 points. Each touchdown was worth 7 points, so how many touchdowns did they make? Use the algebraic equation below to solve the problem.

    $7 = \frac{56}{x}$ $\quad 7x = 56 \quad$ $x = \frac{56}{7}$ $\quad x = 8 \text{ touchdowns}$

12. Keith is writing an essay for a contest. Each mistake is worth $-\frac{1}{2}$ points. If his total is $-5\frac{1}{2}$ points or more, he will win. How many mistakes can he make and still win? Use the algebraic equation below to solve the problem.

    $-\frac{1}{2} = \frac{-5\frac{1}{2}}{x}$ $\quad -\frac{1}{2}x = -5\frac{1}{2} \quad$ $x = -\frac{11}{2} \cdot -\frac{2}{1}$ $\quad x = \frac{22}{2}$ $\quad x = 11$

ANSWERS: WORKSHEET 3-4

1. $93 - x = 47$
   $93 = 47 + x$
   $93 - 47 = x$
   $46 = x$

2. $x + 75 = -5$
   $x = -5 - 75$
   $x = -80$

3. $-x + 10 = 5$
   $10 = 5 + x$
   $10 - 5 = x$
   $5 = x$

4. $-5 + x = 11$
   $x = 11 + 5$
   $x = 16$

5. $x - (-23) = -43$
   $x = -43 + (-23)$
   $x = -66$

6. $-17 + x = -34$
   $x = -34 + 17$
   $x = -17$

7. $-3x = 9$
   $x = -3$

8. $12x = -48$
   $x = -4$

9. $99 = 11x$
   $x = 9$

10. $5x = 105$
    $x = 21$

11. $16x = 176$
    $x = 11$

12. $9x + 5 = 50$
    $9x = 50 - 5$
    $9x = 45$
    $x = 5$

13. $11x - 6 = 115$
    $11x = 115 + 6$
    $11x = 121$
    $x = 11$

14. $8 + 15a = 38$
    $15a = 38 - 8$
    $15a = 30$
    $a = 2$

15. $-33 - 14x = -145$
    $-14x = -145 + 33$
    $-14x = -112$
    $x = 8$

16. $45x - -2 = 92$
    $45x + 2 = 92$
    $45x = 92 - 2$
    $x = 2$

17. $\frac{1}{5}x + 7 = 11$
    $\frac{1}{5}x = 11 - 7$
    $\frac{1}{5}x = 4$
    $x = \frac{4}{1} \div \frac{1}{5}$
    $x = 20$

18. $x - \frac{5}{8} = 2\frac{3}{8}$
    $x = 2\frac{3}{8} + \frac{5}{8}$
    $x = 3$

ANSWERS: WORKSHEET 3-5

1. $\frac{54}{x} = 9$
   $x = 6$

2. $\frac{x}{8} = 7$
   $x = 56$

3. $\frac{48}{x} = 8$
   $x = 6$

4. $\frac{x}{5} = 4$
   $x = 20$

5. $\frac{144}{x} = 12$
   $x = 12$

6. $\frac{56}{x} = 28$
   $x = 2$

7. $\frac{28}{x} + 3 = 10$
   $\frac{28}{x} = 10 - 3$
   $\frac{28}{x} = 7$
   $x = 4$

8. $8 + \frac{56}{x} = 15$
   $\frac{56}{x} = 15 - 8$
   $\frac{56}{x} = 7$
   $x = 8$

9. $\frac{x}{6} - 2 = 4$
   $\frac{x}{6} = 4 + 2$
   $\frac{x}{6} = 6$
   $x = 36$

10. $\frac{1}{3}x = 2$
    $x = 2 \div \frac{1}{3}$
    $x = 6$

11. $\frac{200}{x} = 1$
    $200 = 1x$
    $200 = x$

12. $\frac{1}{4}x = 4$
    $x = 4 \div \frac{1}{4}$
    $x = 16$

13. $\frac{x}{2} - 2 = 3$
    $\frac{x}{2} = 3 + 2$
    $\frac{x}{2} = 5$
    $x = 5 \cdot 2$
    $x = 10$

14. $\frac{96}{x} + 9 = 17$
    $\frac{96}{x} = 17 - 9$
    $\frac{96}{x} = 8$
    $96 = 8x$
    $\frac{96}{8} = \frac{8x}{8}$
    $12 = x$

15. $\frac{14}{x} = 14$
    $x = 1$ *of course*

ANSWERS: WORKSHEET 3-6

Solve for x in terms of y.

1. $x + y = 24$
   $x = 24 - y$

2. $y + x = 18$
   $x = 18 - y$

3. $x + 8 = y$
   $x = y - 8$

4. $x - 4 = y$
   $x = y + 4$

5. $xy = 30$
   $x = \dfrac{30}{y}$

6. $\dfrac{x}{y} = 15$
   $x = 15y$

7. $17 + x = y$
   $x = y - 17$

8. $\dfrac{y}{x} = 16$
   $y = 16x$
   $\dfrac{y}{16} = x$

9. $\dfrac{x}{y} = 77$
   $x = 77y$

Solve for y in terms of x.

10. $\dfrac{48}{y} = x$
    $48 = xy$
    $\dfrac{48}{x} = y$

11. $\dfrac{x}{y} = 4$
    $x = 4y$
    $\dfrac{x}{4} = y$

12. $\dfrac{144}{y} = x$
    $144 = xy$
    $\dfrac{144}{x} = y$

13. $5y = x$
    $y = \dfrac{x}{5}$

14. $9x + y = 50$
    $y = 50 - 9x$

15. $11x + y = 115$
    $y = 115 - 11x$

16. $x = 100y$
    $\dfrac{x}{100} = y$

17. $x = 41y$
    $\dfrac{x}{41} = y$

18. $x = y - 10$
    $x + 10 = y$

ANSWERS: WORKSHEET 3-7A

Write each ratio as a fraction.

1. The ratio of 3 to 7. $\dfrac{3}{7}$

2. The ratio of 5 to 10. $\dfrac{5}{10}$

3. The ratio of 8 to 3. $\dfrac{8}{3}$

4. The ratio of ½ to 3. $\dfrac{\frac{1}{2}}{3}$

5. A fisherman needs 12 pounds of weight for every 60 feet of fishing line. What is the ratio of weight to feet? Reduce your answer.

$$\frac{12\ pounds}{60\ feet} = \frac{1}{5}$$

6. At the school dance, there were 40 boys and 20 girls. What is the ratio of boys to girls? Reduce your answer.

$$\frac{40\ boys}{20\ girls} = \frac{2}{1}$$

7. The votes were counted. There were 6 "yes" votes and 18 "no" votes. What is the ratio of yes to no votes? Reduce your answer.

$$\frac{6}{18} = \frac{1}{3}$$

ANSWERS: WORKSHEET 3-7A page 2

8. The campers received one tent per four campers. What is the ratio of campers to tents? Write your answer in this format: 9:3.

$$4:1$$

9. A race car travels three miles in one minute. Another car drove one mile in one minute. Write a ratio that compares the distance of the race car to the other car.

$$\frac{3 \text{ miles per minute}}{1 \text{ mile per minute}} = \frac{3}{1}$$

10. A recipe for potato salad suggests using 3 potatoes for each serving. What is the ratio of potatoes per person?

$$\frac{3}{1}$$

ANSWERS: WORKSHEET 3-7B

1. Write a ratio that shows 14 dog bones for 7 dogs. Then show how many dog bones per dog.

$$\frac{14}{7} = 2 \; dog \; bones \; per \; dog$$

2. Write a ratio to help you calculate the cost per can of pop. A 6-pack of pop costs $2.94. How much per can? Hint: your answer will be in cents, so your ratio should be in cents not dollars.

$$\frac{294}{6} = 49 \; cents \; per \; pop$$

3. Write a ratio, using the ratio symbol ":" to show 2 teachers to 44 students. **2:44 Reduces down to 1:22**

4. Write two ratios to help you figure out which is the better value. You can spend $249 for 100 t-shirts. Or you can spend $49 to buy 10 t-shirts. Find the price per t-shirt. Which is the better deal?

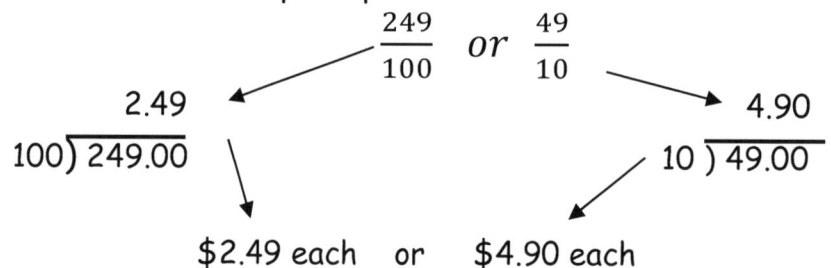

$$\frac{249}{100} \quad or \quad \frac{49}{10}$$

```
   2.49                    4.90
100) 249.00            10 ) 49.00
```

$2.49 each   or $4.90 each

Buying the shirts 100 at a time is the better deal.

5. We traveled 444 miles in 3 days. How many miles did we travel per day?

$$\frac{444}{3} = 148 \; miles \; per \; day$$

6. Write three different phrases to describe this: $\frac{1}{3}$

One third   One divided by three   One to three

ANSWERS: WORKSHEET 3-7B page 2

7. It cost $46.05 for 15 gallons of gas. Write a ratio and then solve it to find the price per gallon.

$$\frac{46.05}{15} = \$3.07 \; per \; gallon$$

8. Pat ran 3 miles in 20 minutes. How many miles did she run per minute?

$$\frac{3}{20} = .15 \; miles \; per \; minute$$

9. Florence spent $11.60 on 5 pounds of hamburger. How much does it cost for 1 pound of hamburger?

$$\frac{11.60}{5} = \$2.32 \; per \; pound$$

ANSWERS: WORKSHEET 3-8A

1. $\dfrac{40}{x} = \dfrac{5}{1}$ $\qquad$ $40 = 5x$
   $\qquad\qquad\qquad\quad\;\; 8 = x$

2. $\dfrac{x}{16} = \dfrac{2}{8}$ $\qquad$ $32 = 8x$
   $\qquad\qquad\qquad\quad\;\; 4 = x$

3. $\dfrac{54}{x} = \dfrac{6}{1}$ $\qquad$ $54 = 6x$
   $\qquad\qquad\qquad\quad\;\; 9 = x$

4. $\dfrac{6}{36} = \dfrac{3}{x}$ $\qquad$ $108 = 6x$
   $\qquad\qquad\qquad\quad\;\; 18 = x$

5. $\dfrac{24}{12} = \dfrac{x}{2}$ $\qquad$ $48 = 12x$
   $\qquad\qquad\qquad\quad\;\; 4 = x$

6. $\dfrac{105}{5} = \dfrac{42}{x}$ $\qquad$ $210 = 105x$
   $\qquad\qquad\qquad\quad\;\; 2 = x$

7. $\dfrac{\frac{2}{5}}{x} = \dfrac{10}{50}$ $\qquad$ $\dfrac{2}{5} \cdot \dfrac{50}{1} = 10x$
   $\qquad\qquad\qquad\quad\;\; \dfrac{100}{5} = 10x$
   $\qquad\qquad\qquad\quad\;\; 20 = 10x$
   $\qquad\qquad\qquad\quad\;\; 2 = x$

8. 7 is to 21 as x is to 6. $\quad \dfrac{7}{21} = \dfrac{x}{6}$ $\qquad$ $42 = 21x$ $\qquad$ $x = 2$

9. 15 is to x as 30 is to 6. $\quad \dfrac{15}{x} = \dfrac{30}{6}$ $\qquad$ $90 = 30x$ $\qquad$ $x = 3$

10. x is to 100 as 9 is to 30. $\quad \dfrac{x}{100} = \dfrac{9}{30}$ $\qquad$ $900 = 30x$ $\qquad$ $x = 30$

11. 7 is to 2 as 14 is to x. $\quad \dfrac{7}{2} = \dfrac{14}{x}$ $\qquad$ $28 = 7x$ $\qquad$ $x = 4$

ANSWERS: WORKSHEET 3-8B

1. There are 3 feet in 1 yard. How many yards are in 120 feet?

$$\frac{3\ ft}{1\ yd} = \frac{120\ ft}{x}$$

$$3x = 120$$

$$x = 40\ yards$$

2. There are 12 inches in 1 foot. How many inches are in 8 feet?

$$\frac{12\ in}{1\ ft} = \frac{x}{8\ ft}$$

$$x = 96\ inches$$

3. There are 16 ounces in 1 pound. How many pounds in 128 ounces?

$$\frac{16\ oz}{1\ lb} = \frac{128\ oz}{x}$$

$$16x = 128$$

$$x = 8\ pounds$$

4. There are 3.28 feet in 1 meter. How many feet in 3 meters?

$$\frac{3.28\ ft}{1\ meter} = \frac{x}{3\ meters}$$

$$x = 9.84\ feet$$

5. One gram equals .035 ounces. 14 grams equal how many ounces?

$$\frac{1\ g}{.035\ oz} = \frac{14\ g}{x}$$

$$x = .49\ ounces$$

ANSWERS: WORKSHEET 3-8B page 2

6. There are 2.54 centimeters in 1 inch. 5 inches equal how many centimeters?

$$\frac{2.54\ cm}{1\ in} = \frac{x}{5\ in}$$

$$x = 12.7\ centimeters$$

7. There are 8 ounces in 1 cup and 16 cups in 1 gallon. How many ounces are in 1 gallon?

$$\frac{8\ oz}{1\ c} = \frac{x}{16\ c}$$

$$x = 128\ ounces$$

ANSWERS: WORKSHEET 3-9

1. $6^2 = 36$
2. $7^2 = 49$
3. $8^2 = 64$

4. $3^2 = 9$
5. $2^2 = 4$
6. $9^2 = 81$

7. $10^2 = 100$
8. $5^2 = 25$
9. $4^2 = 16$

10. $11^2 = 121$
11. $1^2 = 1$
12. $12^2 = 144$

13. $3^2 + 3^2 = 18$
14. $2^2 + 4^2 = 20$
15. $5^2 + 3^2 = 34$

16. $6^2 + 4^2 = 52$
17. $8^2 + 7^2 = 113$
18. $3^2 + 2^2 = 13$

19. $4^2 \cdot 2 = 32$
20. $5^2 \cdot 4 = 100$
21. $6^2 \cdot x = 36x$

22. $6^2 \cdot \frac{1}{2} = 18$
23. $4^2 \div \frac{1}{2} = 32$
24. $2^2 \cdot 3^2 = 36$

ANSWERS: WORKSHEET 3-10

1. $6^3 = 216$
2. $7^3 = 343$
3. $8^3 = 512$

4. $3^3 = 27$
5. $2^3 = 8$
6. $9^3 = 729$

7. $10^3 = 1000$
8. $5^3 = 125$
9. $4^3 = 64$

10. $11^3 = 1,331$
11. $1^3 = 1$
12. $12^3 = 1,728$

13. $3^3 + 3^2 = 36$
14. $2^3 + 4^3 = 72$
15. $5^3 + 3^2 = 134$

16. $6^2 + 4^3 = 100$
17. $8^3 + 10 = 522$
18. $3^3 + 2^3 = 35$

19. $5^3 \cdot 2 = 250$
20. $5^3 \cdot 3 = 375$
21. $4^3 \cdot y = 64y$

22. $2^3 \cdot \frac{1}{4} = 2$
23. $3^3 \div \frac{2}{7} = \frac{189}{2} = 94\frac{1}{2}$

24. $4^3 \cdot 7^2 = 3,136$

ANSWERS: WORKSHEET 3-11

Solve the following.

1. $5^2 = 25$
2. $\sqrt{25} = 5$
3. $6^2 = 36$

4. $\sqrt{36} = 6$
5. $8^2 = 64$
6. $\sqrt{64} = 8$

7. $2^2 = 4$
8. $\sqrt{100} = 10$
9. $9^2 = 81$

10. $\sqrt{49} = 7$
11. $\sqrt{121} = 11$
12. $\sqrt{16} = 4$

13. $\sqrt{144} - \sqrt{16} = 8$
    $12 - 4 = 8$

14. $\sqrt{4} + \sqrt{25} = 7$
    $2 + 5 = 7$

15. $\sqrt{49} \cdot \sqrt{36} = 42$
    $7 \cdot 6 = 42$

16. $\sqrt{64} \cdot 2^2 = 32$
    $8 \cdot 4 = 32$

17. $\sqrt{100} \div 5 = 2$
    $10 \div 5 = 2$

18. $\sqrt{81} \cdot x = 36$
    $9 \cdot x = 36$
    $9x = 36$
    $x = 4$

ANSWERS: CHAPTER 1 REVIEW TEST

1. $29 + x = 104$
    $x = 104 - 29$
    $x = 75$

2. $9x = 63$
    $\frac{9x}{9} = \frac{63}{9}$
    $x = 7$

3. $\frac{x}{4} = 6$
    $\frac{x}{4}(4) = 6 \times 4$
    $x = 24$

4. $\frac{48}{x} = 8$
    $\frac{48}{x}(x) = 8x$
    $48 = 8x$
    $6 = x$

5. $\frac{10}{x} \cdot \frac{14}{1} = 70$
    $\frac{140}{1x} = 70$
    $\frac{140}{x}(x) = 70x$
    $140 = 70x$
    $\frac{140}{70} = \frac{70x}{70}$
    $2 = x$

6. $52 - x = 10$
    $-x = 10 - 52$
    $-x = -42$
    $x = 42$

7. $-201 - 3x = -225$
    $-3x = -225 + 201$
    $-3x = -24$
    $\frac{-3x}{-3} = \frac{-24}{-3}$
    $x = 8$

8. $\frac{1}{2}x + 12 = 10$
    $\frac{1}{2}x = 10 - 12$
    $\frac{1}{2}x = -2$
    $x = -2 \div \frac{1}{2}$
    $x = -4$

9. $\frac{11}{x} = 11$
    $\frac{11}{x}(x) = 11x$
    $11 = 11x$
    $1 = x$

10. $\frac{48}{x} + 12 = 24$
    $\frac{48}{x} = 24 - 12$
    $\frac{48}{x} = 12$
    $x = 4$

11. $\frac{9}{x} + 6 = 9$
    $\frac{9}{x} = 9 - 6$
    $\frac{9}{x} = 3$
    $9 = 3x$
    $3 = x$

12. $\frac{3}{5}x - 12 = -2\frac{2}{5}$
    $\frac{3}{5}x = -2\frac{2}{5} + 12$
    $\frac{3}{5}x = 9\frac{3}{5}$
    $x = 9\frac{3}{5} \div \frac{3}{5} =$
    $x = \frac{48}{5} \times \frac{5}{3}$
    $x = 16$

Solve for x in the following ratios.

13. $\frac{5}{x} = \frac{10}{12}$
    $60 = 10x$
    $6 = x$

14. $\frac{x}{32} = \frac{2}{8}$
    $64 = 8x$
    $8 = x$

15. $\frac{72}{9} = \frac{8}{x}$
    $72 = 72x$
    $1 = x$

ANSWERS: CHAPTER 1 REVIEW TEST page 2

Solve for x in terms of y.

16. $x + y = 325$
    $x = 325 - y$

17. $\frac{x}{8} = y$
    $x = 8y$

18. $x - 2 = y$
    $x = y + 2$

19. $6^2 + 4^2 =$
    $36 + 16 = 52$

20. $8^2 \cdot 10^2 =$
    $64 \cdot 100 = 6400$

21. $\sqrt{81} \cdot \sqrt{36} =$
    $9 \times 6 = 54$

22. $5^3 \cdot y = 125y$

23. $2^3 \div \frac{1}{6} =$
    $\frac{8}{1} x \frac{6}{1} = 48$

24. $\sqrt{100} \cdot 7^2 =$
    $10 \times 49 = 490$

ANSWERS: WORKSHEET 3-12

Name each of the following as a term, an expression, or an equation.

1. $3x$
   Term

2. $4x - 2x + 10$
   Expression

3. $5y^2$
   Term

4. $5y + x = 40$
   Equation

5. $-10 + 2y - 7y + 8$
   Expression

6. $\sqrt{64}$
   Term

Solve each of the following terms.

7. $\sqrt{81} = 9$

8. $3^3 = 27$

9. If $x = 4$, then $3x = 12$

Solve the following expressions.

10. $3^2 + 7 = 16$
    $9 + 7 = 16$

11. $\sqrt{36} - 2^2 = 2$
    $6 - 4 = 2$

12. If $x = 5$, then $2x + 5x =$
    $(2 \cdot 5) + (5 \cdot 5) =$
    $10 + 25 = 35$

Solve for x in the following equations.

13. $\sqrt{x} = 5$
    $x = 5^2$
    $x = 25$

14. $7x + 3 = 31$
    $7x = 31 - 3$
    $7x = 28$
    $\frac{7x}{7} = \frac{28}{7}$
    $x = 4$

15. $45 = 9x + 9$
    $45 - 9 = 9x$
    $36 = 9x$
    $\frac{36}{9} = \frac{9x}{9}$
    $4 = x$

Solve for m in the following equations.

16. $8m + 2 = 58$
    $8m = 58 - 2$
    $8m = 56$
    $\frac{8m}{8} = \frac{56}{8}$
    $m = 7$

17. $\frac{m}{8} = 8$
    $m = 8 \cdot 8$
    $m = 64$

18. $m^2 = 7$
    $\sqrt{m^2} = \sqrt{7}$
    $m = \sqrt{7}$

ANSWERS: WORKSHEET 3-13

Simplify each of the following expressions by combining like terms.

1. $3y + 4y - 2x =$
    $7y - 2x$

2. $4x - 2x + 3y - 2y =$
    $2x + y$

3. $7y - 2x + 4y =$
    $11y - 2x$

4. $6x - 5x + 8y + 2x =$
    $3x + 8y$

5. $8m - 2m + 3r =$
    $6m + 3r$

6. $7x + 7y + 7m + 3x =$
    $10x + 7y + 7m$

7. $4x^2 + 3x^2 =$
    $7x^2$

8. $8rst + 9rst + rst =$
    $18rst$

9. $5 + 7b^2 - 2 - 3b^2 =$
    $3 + 4b^2$

10. $5xy + 7xy + 17xy^2 =$
    $12xy + 17xy^2$

11. $3mn - mn =$
    $2mn$

12. $6xy^2 + 5xy^2 + 8xy^3 =$
    $11xy^2 + 8xy^3$

13. $18y + 2 =$
    $18y + 2$

14. $27rst + 33rst - 7rst + 4rs =$
    $53rst + 4rs$

ANSWERS: WORKSHEET 3-14

1. $x^2 \cdot x^2 = x^4$ 2. $x^3 \cdot x^4 = x^7$ 3. $y^6 \cdot y^5 = y^{11}$

4. $xy^2 \cdot xy = x^2y^3$ 5. $abc^2 \cdot ab \cdot abc = a^3b^3c^3$ 6. $xyz \cdot x^4y^3z = x^5y^4z^2$

7. $4x^2 \cdot 3x^3 = 12x^5$ 8. $7ab^3 \cdot ab^3 = 7a^2b^6$ 9. $9a \cdot 3a = 27a^2$

10. $7x \cdot x^3 = 7x^4$ 11. $x \cdot y = xy$ 12. $5x^3 \cdot 7y^4 = 35x^3y^4$

13. $11y^2 \cdot 11xy^2 = 121xy^4$ 14. $x^{10} \cdot x^{10} = x^{20}$ 15. $9abc^2 \cdot 3xyz^2 = 27abc^2xyz^2$

16. $14a^2 \cdot a = 14a^3$ 17. $x \cdot y \cdot z \cdot z = xyz^2$ 18. $a^2b^2c^2 \cdot abc = a^3b^3c^3$

19. $\frac{3}{5}x \cdot x = \frac{3}{5}x^2$ 20. $8^2 \cdot y^2 = 64y^2$

21. $21x^2 \cdot x^2 + 3y = 21x^4 + 3y$

ANSWERS: WORKSHEET 3-15

Solve for x.

1. $25x = (63 + 37)$
    $25x = (100)$
    $\frac{25x}{25} = \frac{100}{25}$
    $x = 4$

2. $52 + x = (33 \cdot 3)$
    $52 + x = (99)$
    $x = 99 - 52$
    $x = 47$

3. $x + (13 - 4) = 15$
    $x + (9) = 15$
    $x = 15 - 9$
    $x = 6$

4. $(7 \cdot 3) + x = 126$
    $(21) + x = 126$
    $x = 126 - 21$
    $x = 105$

5. $x + (5 \cdot 5) = 45$
    $x + (25) = 45$
    $x = 45 - 25$
    $x = 20$

6. $6x = (14 + 16) \cdot 2$
    $6x = (30) \cdot 2$
    $6x = 60$
    $\frac{6x}{6} = \frac{60}{6}$
    $x = 10$

7. $(15 \div 3) + x^2 = 54$
    $(5) + x^2 = 54$
    $x^2 = 54 - 5$
    $x^2 = 49$
    $\sqrt{x^2} = \sqrt{49}$
    $x = 7$

8. $\sqrt{25} + (3 \cdot 5) = 2x$
    $5 + (15) = 2x$
    $20 = 2x$
    $\frac{20}{2} = \frac{2x}{2}$
    $10 = x$

9. $\sqrt{x} + (2^2 \cdot 3^2) = 45$
    $\sqrt{x} + (4 \cdot 9) = 45$
    $\sqrt{x} + (36) = 45$
    $\sqrt{x} = 45 - 36$
    $\sqrt{x} = 9$
    $(\sqrt{x})^2 = 9^2$
    $x = 81$

10. $(x \cdot 3) + 2 = 29$
    $(3x) + 2 = 29$
    $3x = 29 - 2$
    $3x = 27$
    $x = 9$

ANSWERS: WORKSHEET 3-15 page 2

11. $(12 \div 4) \cdot 2x = 54$
    $(3) \cdot 2x = 54$
    $6x = 54$
    $\frac{6x}{6} = \frac{54}{6}$
    $x = 9$

12. $(-5 \cdot -5) \div x = 5$
    $(25) \div x = 5$
    $25 = 5x$
    $\frac{25}{5} = \frac{5x}{5}$
    $5 = x$

13. $\left(\frac{1}{4} + \frac{12}{16}\right) \cdot 2x + 3 = 5$
    $\left(\frac{4}{16} + \frac{12}{16}\right) \cdot 2x + 3 = 5$
    $1 \cdot 2x + 3 = 5$
    $2x = 5 - 3$
    $2x = 2$
    $\frac{2x}{2} = \frac{2}{2}$
    $x = 1$

14. $(x \cdot x) + 4 = 53$
    $(x^2) + 4 = 53$
    $x^2 = 53 - 4$
    $x^2 = 49$
    $\sqrt{x^2} = \sqrt{49}$
    $x = 7$

15. $[(12 + 14) \div 2] + 2x = 27$
    $(26 \div 2) + 2x = 27$
    $13 + 2x = 27$
    $2x = 27 - 13$
    $2x = 14$
    $\frac{2x}{2} = \frac{14}{2}$
    $x = 7$

16. $[(7 \cdot 3) + 4] - 5 = 4x$
    $(21 + 4) - 5 = 4x$
    $25 - 5 = 4x$
    $20 = 4x$
    $\frac{20}{4} = \frac{4x}{4}$
    $5 = x$

ANSWERS: WORKSHEET 3-16

Solve for x in the equations below. Be sure to go in the proper order.

1. $2^2 - x + (3 \cdot 2) = 8$
   $4 - x + (6) = 8$
   $4 - x = 8 - 6$
   $4 - x = 2$
   $-x = 2 - 4$
   $-x = -2 \quad x = 2$

2. $(4 \cdot 5) - 4^2 + x = 24$
   $(20) - 16 + x = 24$
   $4 + x = 24$
   $x = 24 - 4$
   $x = 20$

3. $3^2 + (8 \cdot 3) = 3x$
   $9 + (24) = 3x$
   $33 = 3x$
   $11 = x$

4. $7x + 5x - (4 \cdot 2x) = 10^2$
   $7x + 5x - (8x) = 100$
   $4x = 100$
   $x = 25$

5. $2^3 - (6 \cdot x) = -4$
   $8 - (6x) = -4$
   $-6x = -4 - 8$
   $-6x = -12$
   $x = 2$

6. $\frac{(9^2 + 3^2)}{x} = 9$
   $\frac{(81+9)}{x} = 9$
   $\frac{90}{x} = 9$
   $90 = 9x$
   $10 = x$

7. $\sqrt{9} + 2^2 - [20 - (6 \cdot 3)] =$
   $3 + 4 - (20 - 18) =$
   $7 - 2 = 5$

8. $(7 - 5) \cdot 3^2 \cdot \sqrt{4} =$
   $(2) \cdot 9 \cdot 2 = 36$

9. $\left(\frac{1}{2} + \frac{3}{8}\right) \cdot \left(\frac{2}{3} - \frac{1}{6}\right) \cdot 2^2 =$
   $\left(\frac{4}{8} + \frac{3}{8}\right) \cdot \left(\frac{4}{6} - \frac{1}{6}\right) \cdot 4 =$
   $\frac{7}{8} \cdot \frac{3}{6} \cdot \frac{4}{1} = \frac{84}{48} = 1\frac{36}{48} = 1\frac{3}{4}$

10. $\frac{\sqrt{100} \cdot 6^2}{2} =$
    $\frac{10 \cdot 36}{2} = \frac{360}{2} = 180$

ANSWERS: WORKSHEET 3-17

1. $\sqrt{81} \cdot 4 + (\sqrt{4} + \sqrt{16}) =$
   $\sqrt{81} \cdot 4 + (2 + 4) =$
   $9 \cdot 4 + (6) =$
   $36 + 6 = 42$

2. $10^2 - 2 \cdot 5^2 + 2x =$
   $100 - 2 \cdot 25 + 2x =$
   $100 - 50 + 2x =$
   $50 + 2x$

3. $3x + 2x - [(6^2 + 4) \cdot 2] =$
   $3x + 2x - [(36 + 4) \cdot 2] =$
   $3x + 2x - [40 \cdot 2] =$
   $3x + 2x - [80] =$
   $5x - 80$

4. $4 \cdot (3 + 5) - \sqrt{49} \cdot \sqrt{4} =$
   $4 \cdot (8) - \sqrt{49} \cdot \sqrt{4} =$
   $4 \cdot (8) - 7 \cdot 2 =$
   $32 - 14 = 18$

5. $3x + 7^2 - \sqrt{25} = 74$
   $3x + 49 - 5 = 74$
   $3x + 44 = 74$
   $3x = 74 - 44$
   $3x = 30$
   $\frac{3x}{3} = \frac{30}{3}$
   $x = 10$

6. $56 - 5x = 11$
   $-5x = 11 - 56$
   $-5x = -45$
   $\frac{-5x}{5} = \frac{-45}{5}$
   $-x = -9$
   $x = 9$

7. $3^2 - x + (3 \cdot 2) = 11$
   $3^2 - x + (6) = 11$
   $9 - x + (6) = 11$
   $9 - x = 11 - 6$
   $9 - x = 5$
   $-x = 5 - 9$
   $-x = -4$, so $x = 4$

8. $(x \cdot x) + 5^2 = 125$
   $(x^2) + 25 = 125$
   $x^2 = 125 - 25$
   $x^2 = 100$
   $\sqrt{x^2} = \sqrt{100}$
   $x = 10$

9. $3x - 7x + x + 2y$
   $-3x + 2y$

10. $4x^2 \cdot 5x^2 + 2x + 6x$
    $20x^4 + 8x$

11. $6y \cdot 5y^2 - 6x + 7y^3$
    $30y^3 - 6x + 7y^3 =$
    $37y^3 - 6x$

12. $3y^3 \cdot 2y^3 + 5y^6$
    $6y^6 + 5y^6 = 11y^6$

ANSWERS: WORKSHEET 3-18

Solve the following by using the distributive property of multiplication.

1. $2(6 + 9) =$
   $(2 \cdot 6) + (2 \cdot 9) =$
   $12 + 18 = 30$

2. $3x(2x - 3) =$
   $(3x \cdot 2x) - (3x \cdot 3) =$
   $6x^2 - 9x$

3. $5x^2(2x + 5x^2) =$
   $(5x^2 \cdot 2x) + (5x^2 \cdot 5x^2) =$
   $10x^3 + 25x^4$

4. $20x^3(3x - 7x^2) =$
   $(20x^3 \cdot 3x) - (20x^3 \cdot 7x^2) =$
   $60x^4 - 140x^5$

5. $3x(y - x) =$
   $(3x \cdot y) - (3x \cdot x) =$
   $3xy - 3x^2$

6. $9^2(x + y) =$
   $(81 \cdot x) + (81 \cdot y) =$
   $81x + 81y$

7. $-4(2x - 3) =$
   $(-4 \cdot 2x) - (-4 \cdot 3) =$
   $-8x + 12$

8. $-3a^2(a + b) =$
   $(-3a^2 \cdot a) + (-3a^2 \cdot b) =$
   $-3a^3 - 3a^2b$

9. $m(x + y) =$
   $(m \cdot x) + (m \cdot y)$
   $mx + my$

10. $5rs(6rs - rs) =$
    $(5rs \cdot 6rs) - (5rs \cdot rs) =$
    $30r^2s^2 - 5r^2s^2 = 25r^2s^2$

11. $ab(3a^2b - 4ab^2) =$
    $(ab \cdot 3a^2b) - (ab \cdot 4ab^2) =$
    $3a^3b^2 - 4a^2b^3$

12. $\frac{1}{2}(a + b) =$
    $\left(\frac{1}{2} \cdot a\right) + \left(\frac{1}{2} \cdot b\right) =$
    $\frac{1}{2}a + \frac{1}{2}b$

13. $-5x(x^3 - 2x^2) =$
    $(-5x \cdot x^3) - (-5x \cdot 2x^2) =$
    $-5x^4 + 10x^3$

14. $-4a(3a - 2b) =$
    $(-4a \cdot 3a) - (-4a \cdot 2b) =$
    $-12a^2 + 8ab$

ANSWERS: WORKSHEET 3-19

1. $2(5 - x) =$
   $\mathbf{10 - 2x}$

2. $3x(6x + 4) =$
   $\mathbf{18x^2 + 12x}$

3. $2x(2x + 3) =$
   $\mathbf{4x^2 + 6x}$

4. $-x(y - 2) =$
   $\mathbf{-xy + 2x}$

5. $4x + 2x = 48$
   $6x = 48$
   $\frac{6x}{6} = \frac{48}{6}$
   $x = 8$

6. $3^3 - 4x + (6 \cdot 5) = 41$
   $27 - 4x + (30) = 41$
   $27 + 30 - 4x = 41$
   $57 - 4x = 41$
   $-4x = 41 - 57$
   $-4x = -16$
   $\frac{-4x}{-4} = \frac{-16}{-4}$
   $x = 4$

7. $3x(8 + 3) - (4^2) = 83$
   $(24x + 9x) - 16 = 83$
   $33x - 16 = 83$
   $33x = 83 + 16$
   $33x = 99$
   $x = 3$

8. $\sqrt{x} = (7 \cdot 2^2) - (7 \cdot \sqrt{9})$
   $\sqrt{x} = (28) - (21)$
   $\sqrt{x} = 7$
   $(\sqrt{x})^2 = 7^2$
   $x = 49$

9. $(48 - 4) \div 2^2 + 2 \cdot 2^3 =$
   $44 \div 4 + 2 \cdot 8 =$
   $11 + 16 = 27$

10. $(-20 + 4) \div 2^3 - 3 \cdot 4 =$
    $-16 \div 8 - 12 =$
    $-2 - 12 = -14$

11. $7 - x = y$
    $7 = y + x$
    $7 - y = x$

12. $x + y = 20$
    $x = 20 - y$

13. $x \div 5 = y$
    $x = 5y$

14. $\frac{36}{x} = y$
    $x\left(\frac{36}{x}\right) = xy$
    $36 = xy$
    $\frac{36}{y} = \frac{xy}{y}$ $\quad \frac{36}{y} = x$

ANSWERS: WORKSHEET 3-19A

1. $2x(3y - 4x) + 3y(6x + 9y) =$

   $6xy - 8x^2 + 18xy + 27y^2$

   $\mathbf{24xy - 8x^2 + 27y^2}$

2. $5a(7a - 6b) - 3b(4a - 2b) =$

   $35a^2 - 30ab - 12ab + 6b^2$

   $\mathbf{35a^2 - 42ab + 6b^2}$

3. $-5x(2x - 8y) - 7x(4x + 9y) =$

   $-10x^2 + 40xy - 28x^2 - 63xy$

   $\mathbf{-38x^2 - 23xy}$

4. $-x(x + y) + y(y - x) =$

   $-x^2 - xy + y^2 - xy$

   $\mathbf{-x^2 - 2xy + y^2}$

5. $-.35a(-1.2a + 3b) - .14a(3.1b - 6.2a) =$

   $.42a^2 - 1.05ab - .434ab + .868a^2$

   $\mathbf{1.288a^2 - 1.484ab}$

6. $\frac{3}{8}x\left(\frac{1}{2}y - \frac{1}{4}x\right) - \frac{7}{16}x\left(\frac{5}{8}y + \frac{1}{8}x^2\right) =$

   $\frac{3}{16}xy - \frac{3}{32}x^2 - \frac{35}{128}xy - \frac{7}{128}x^3$

   $\frac{3}{16}xy - \frac{35}{128}xy = \quad \frac{24}{128}xy - \frac{35}{128}xy = -\frac{11}{128}xy$

   $-\frac{11}{128}xy - \frac{3}{32}x^2 - \frac{7}{128}x^3$

ANSWERS: WORKSHEET 3-19A Page 2

7. $14mn(mn + 8n) - 8m(4n - 6n^2) =$

       $14m^2n^2 + 112mn^2 - 32mn + 48mn^2$

       $\mathbf{14m^2n^2 + 160mn^2 - 32mn}$

8. $.5x(2xy - 4.2y) + .25y(.6x - .3x^2) =$

       $1x^2y - 2.1xy + .15xy - .075x^2y$

       $1x^2y - .075x^2y - 2.1xy + .15xy$

       $\mathbf{.925x^2y - 1.95xy}$

9. $250g(30g - 100h) + 440g(g + h) =$

       $7{,}500g^2 - 25{,}000gh + 440g^2 + 440gh$

       $7{,}500g^2 + 440g^2 - 25{,}000gh + 440gh$

       $\mathbf{7{,}940g^2 - 24{,}560gh}$

10. $a^2b(bc^2 - abc^2) - a^3(4b + 3b^2c^2) =$

       $a^2b^2c^2 - a^3b^2c^2 - 4a^3b - 3a^3b^2c^2 =$

       $a^2b^2c^2 - a^3b^2c^2 - 3a^3b^2c^2 - 4a^3b =$

       $\mathbf{a^2b^2c^2 - 4a^3b^2c^2 - 4a^3b}$

ANSWERS: CHAPTER 2 REVIEW TEST

1. Which one of the following is a term?

    (3xy)   4y - 2y = 6y $8y^2 + 3x^3$

2. Which one of the following is an expression?

    4abc   4(x + 2) = 36   (3ab + $4ab^2$)

3. Which one of the following is an equation?

    (3 + 5 = 8)   4x - 3x   3xyz

4. If x = 7, then what is: 8x $8 \times 7 = 56$

5. If x = 6, then what is: $\dfrac{42}{x}$ $42 \div 6 = 7$

6. Combine the like terms: $2xy^4 - 4x + 6xy^4 + x - 3y + xy^4 =$

    $9xy^4 - 3x - 3y$

7. Solve for m. $\sqrt{m} = 16$

    $(\sqrt{m})^2 = 16^2$ $m = 256$

8. $xy^3 \cdot x^2 \cdot y^4 = \boldsymbol{x^3 y^7}$

9. $4a \cdot 6a^2 \cdot b^3 \cdot b = \boldsymbol{24a^3 b^4}$

10. Simplify the following: $3(2a + 4b) =$ $\boldsymbol{6a + 12b}$

11. Solve for x: $(x \cdot 4) + 5 = 17$

    $4x + 5 = 17$
    $4x = 17 - 5$
    $4x = 12$
    $x = 3$

ANSWERS: CHAPTER 2 REVIEW TEST page 2

12. Solve for x: $3(9 + x) = 45$

    $27 + 3x = 45$
    $3x = 45 - 27$
    $3x = 18$
    $x = 6$

13. Write out the proper order of operations below:
    Parenthesis, Exponents, Multiplication, Division, Addition, Subtraction

14. $[(9 \cdot 5) + 5] - 3^2 \cdot 6 + 4 =$
    $[(45) + 5] - 9 \cdot 6 + 4 =$
    $50 - 54 + 4 =$
    $-4 + 4 = 0$

15. $(5^2 + 5 - 2^2) - \sqrt{81} \cdot 3 - 7 =$
    $(25 + 5 - 4) - 9 \cdot 3 - 7 =$
    $(26) - 27 - 7 = -8$

16. $\frac{(4^2 + 3^2)}{x} = 5$ $\frac{(25)}{x} = 5$
    $x = 5$

17. $3rs(6r - 4s)$ $18r^2s - 12rs^2$

18. $\frac{1}{2}\left(4a + \frac{5}{7}b\right)$ $2a + \frac{5}{14}b$

19. $4(6x + 3a) + 3(7a - 8x)$ $24x + 12a + 21a - 24x = 33a$

20. $9a(7a + 4a) + a(a - b)$ $63a^2 + 36a^2 + a^2 - ab =$
    $100a^2 - ab$

21. $13a^2(ab + 4b) + 6a(2ab + b)$ $13a^3b + 52a^2b + 12a^2b + 6ab$
    $13a^3b + 64a^2b + 6ab$

22. $-11xy^2(9y - xy) - 5x^4(2y - x)$
    $-99xy^3 + 11x^2y^3 - 10x^4y + 5x^5$

ANSWERS: WORKSHEET 3-20

1. Plot the following coordinate points on the graph below:
   (6, 7) (4, 4) (2, 3) (-3, -2) (-2, -5)

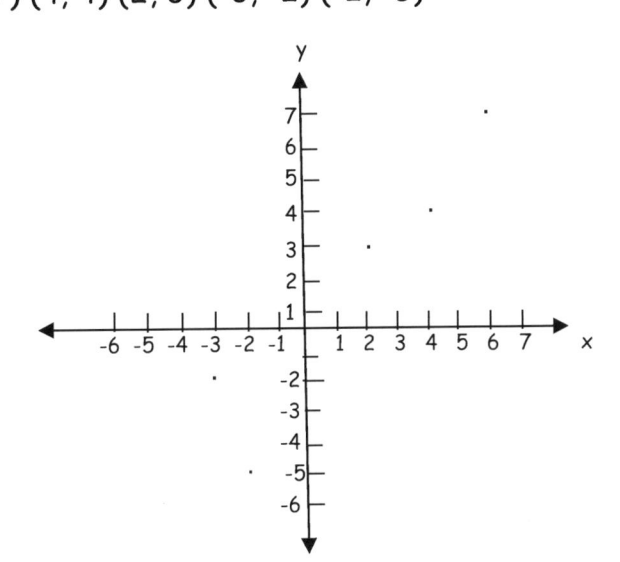

2. Below is a graph. Locate the 4 points and name them (x, y).

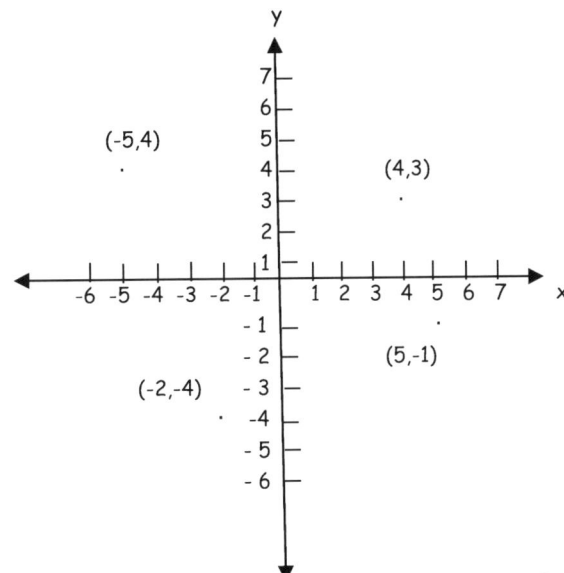

   (-5, 4)   (4, 3)   (5, -1)   (-2, -4)

3. Which number in the coordinates (6, 7) refers to the y axis. Does the y-axis go up and down or left to right? **Y = 7. Up and down.**

ANSWERS: WORKSHEET 3-20 page 2

4. Below is a graph. Plot the following points and then connect the dots to make a straight line. (-4, -4), (-2, -2), (0, 0), (1, 1) and (5, 5).

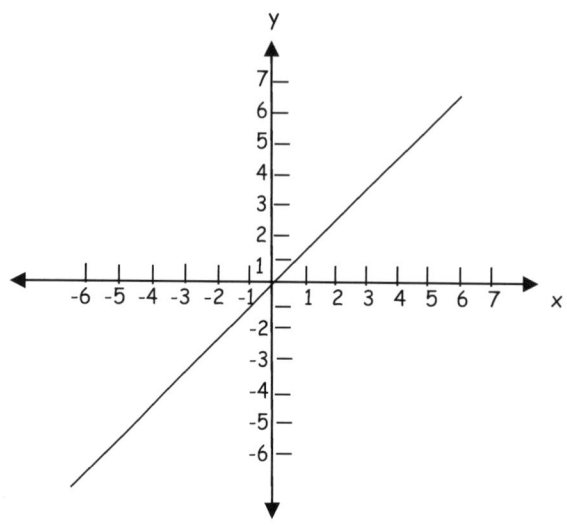

5. Look back at problem number 4. You drew a straight line to connect 5 different points. Name 3 other points that your line passes through.
(-3,-3), (-1,-1), (2,2), (3,3), (4,4) or (-5,-5)

9. If you have two negative coordinates such as (-3, -2) your point will always be located in:
   d. **The lower left side.**

10. "Read X First" means "x" is the "first" number in a set of coordinates and the x axis is the number line that goes in the same direction as we "read."

ANSWERS: WORKSHEET 3-21

Use the following equations to find the missing y values for each table.

1. $y = x + 3$

x	y
1	4
2	5
3	6

2. $y = 3x + 1$

x	y
1	4
2	7
3	10

3. $y = 2x - 2$

x	y
1	0
2	2
3	4

4. $y = 3x - 4$

x	y
1	-1
2	2
3	5

5. $y = 2x - 5$

x	y
1	-3
2	-1
3	1

ANSWERS: WORKSHEET 3-22

Coordinates	Fill in formula	Do the math
(2, 2) and (7, 4)	$m = \frac{4-2}{7-2}$	$m = \frac{2}{5}$
(-2, 4) and (4, 10)	$m = \frac{10-4}{4--2}$	$m = \frac{6}{6}$ or 1
(4, 1) and (10, 12)	$m = \frac{12-1}{10-4}$	$m = \frac{11}{6}$
(3, 2) and (5, 4)	$m = \frac{4-2}{5-3}$	$m = \frac{2}{2}$ or 1
(5, 7) and (3, -3)	$m = \frac{-3-7}{3-5}$	$m = \frac{-10}{-2}$ or 5
(2, 5) and (7, 1)	$m = \frac{1-5}{7-2}$	$m = \frac{-4}{5}$
(-3, 3) and (2, 3)	$m = \frac{3-3}{2--3}$	$m = \frac{0}{5}$ or 0

If the slope = 0, then the line is flat. Plot the points to see for yourself.

| (-4, 2) and (3, -4) | $m = \frac{-4-2}{3--4}$ | $m = \frac{-6}{7}$ |

When the slope is a negative number, it means it is going downhill.

ANSWERS: WORKSHEET 3-22 page 2

The slope of ten different lines are listed below. State which direction the line would go in, if drawn on a graph (uphill, downhill, horizontal or vertical).

8. $m = \frac{8}{5}$ A positive number indicates that the line is going **uphill**.

9. $m = \frac{6}{12}$ A positive number indicates that the line is going **uphill**.

10. $m = -1$ A negative number indicates that the line is going **downhill**.

11. $m = 0$ A slope of zero indicates that the line is **horizontal**.

12. $m = \frac{0}{-5} = 0$ A slope of zero indicates that the line is **horizontal**.

13. $m = 1\frac{1}{4}$ A positive number indicates that the line is going **uphill**.

14. $m = \frac{5}{0}$ $is\ undefined$ This slope is undefined and, therefore, must be **vertical**.

15. $m = \frac{1}{-2}$ A negative number indicates that the line is going **downhill**.

16. $m = 2$ A positive number indicates that the line is going **uphill**.

17. $m = \frac{9}{0}$ $is\ undefined$ This slope is undefined and, therefore, must be **vertical**.

ANSWERS: WORKSHEET 3-23

Coordinate	Slope	Formula filled in	Simplified	Solve for b
(2, 4)	$m = \frac{1}{2}$	$4 = \frac{1}{2} \cdot 2 + b$	$4 = 1 + b$	$b = 3$
(0, 3)	$m = \frac{1}{2}$	$3 = \frac{1}{2} \cdot 0 + B$	$3 = 0 + b$	$b = 3$
(1, 2)	$m = 1$	$2 = 1 \cdot 1 + b$	$2 = 1 + b$	$b = 1$
(2, 3)	$m = 1$	$3 = 1 \cdot 2 + b$	$3 = 2 + b$	$b = 1$
(5, 0)	$m = 4$	$0 = 4 \cdot 5 + b$	$0 = 20 + b$	$b = -20$
(-2, 3)	$m = 0$	$3 = 0 \cdot -2 + b$	$3 = 0 + b$	$b = 3$
(-4, -3)	$m = \frac{2}{3}$	$-3 = \frac{2}{3} \cdot -4 + b$		

Turn -4 into a fraction.	$-3 = \frac{2}{3} \cdot \frac{-4}{1} + b$	
Multiply the fractions.	$-3 = \frac{-8}{3} + b$	
Opposite of $\frac{-8}{3}$ is $+\frac{8}{3}$.	$-3 + \frac{8}{3} = b$	
Turn -3 into a fraction.	$\frac{-3}{1} + \frac{8}{3} = b$	Find a common denominator. $\frac{-9}{3} + \frac{8}{3}$
Add the fractions.	$\frac{-1}{3} = b$	

Coordinate	Slope	Formula filled in	Simplified	Solve for b	
(5, -2)	$m = 2$	$-2 = 2 \cdot 5 + b$	$-2 = 10 + b$	$b = -10 - 2$	$b = -12$

ANSWERS: WORKSHEET 3-24

1. I will give you the slope and the y-intercept for several different lines. You practice filling in the formula, "y = mx + b" with these numbers, so we have an equation for each line. The first one is done for you.

$$y = mx + b$$

Slope	Y-intercept	Equation
m = 2	b = 3	y = 2x + 3
m = 3	b = 1	y = 3x + 1
m = 6	b = 0	y = 6x + 0
m = $\frac{1}{2}$	b = 4	y = $\frac{1}{2}$x + 4
m = 1	b = -3	y = 1x - 3
m = 5	b = -1	y = 5x - 1

2. Fill in the blanks. Write what each letter in the y-intercept formula represents.

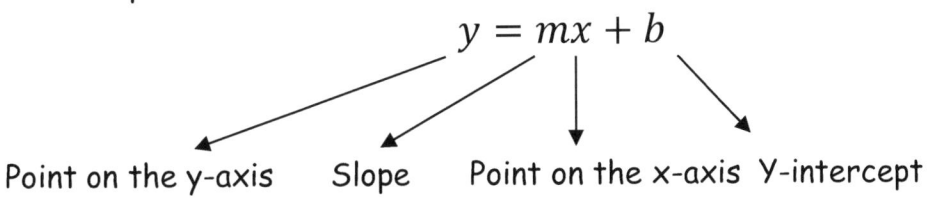

$$y = mx + b$$

Point on the y-axis   Slope   Point on the x-axis   Y-intercept

3. What is the y-intercept?
   The y-intercept is the point where your line crosses the y-axis.

ANSWERS: CHAPTER 3 REVIEW TEST

1. Y = 2x -1

x	y
2	3
3	5
4	7

y = 2 · 2 -1
Y = 3

Y = 2 · 3 -1
Y = 5

y = 2 · 4 - 1
Y = 7

2. Draw a graph and plot the points you found from the first problem. Then connect the dots.

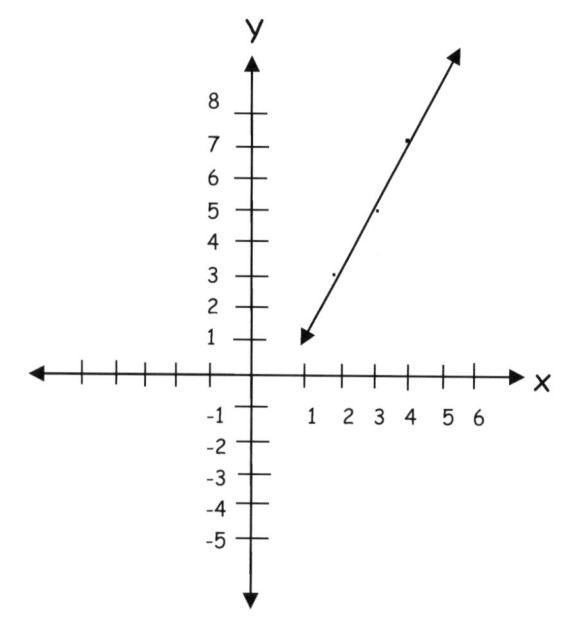

3. Use the slope formula to find the slope of the line you drew on your graph.

$$m = \frac{5-3}{3-2}$$

$$m = \frac{2}{1} = 2$$

ANSWERS: CHAPTER 3 REVIEW TEST page 2

4. Use the formula below to find the y-intercept of your line.

$$y = mx + b$$

3 = 2 · 2 + b
3 = 4 + b
3 − 4 = b
− 1 = b

5. Make an equation for a line that has these two coordinates:
(2, 4) and (4, 8).

First get the slope.

$$m = \frac{y_2 - y_1}{x_2 - x_1}$$

$$m = \frac{8 - 4}{4 - 2}$$

$$m = \frac{4}{2} = 2$$

Next, find the y-intercept, b.

4 = 2 · 2 + b
4 = 4 + b
4 − 4 = b
0 = b   The line intercepts the y-axis at 0; directly in the center of the graph.

Our equation is y = 2x + 0

ANSWERS: FINAL TEST

1. $\frac{1}{2}x \cdot 8 = 32$
   $4x = 32$
   $\frac{4x}{4} = \frac{32}{4}$
   $x = 8$

2. $x + 7^2 = 62$
   $x = 62 - 49$
   $x = 13$

3. $5^3 = x$
   $125 = x$

4. $x(5 + 8) = 78$
   $5x + 8x = 78$
   $13x = 78$
   $x = 6$

5. $5 + 6 \times 9 - x = 51$
   $5 + 54 - x = 51$
   $59 - x = 51$
   $-x = 51 - 59$
   $-x = -8, \ so \ x = 8$

6. $\frac{x}{9} = \frac{4}{6}$
   $36 = 6x$
   $6 = x$

7. $\frac{1}{3}x = \frac{7}{24}$
   $\frac{\frac{1}{3}x}{\frac{1}{3}} = \frac{7}{24} \div \frac{1}{3}$
   $x = \frac{7}{24} \times \frac{3}{1}$
   $x = \frac{7}{8}$

8. $\sqrt{x} = 8$
   $(\sqrt{x})^2 = 8^2$
   $x = 64$

9. $\frac{75}{x} = 15$
   $\left(\frac{75}{x}\right)x = 15x$
   $75 = 15x$
   $\frac{75}{15} = \frac{15x}{15}$
   $5 = x$

10. $x^2 = 81$
    $\sqrt{x^2} = \sqrt{81}$
    $x = 9$

11. $\frac{x}{7} = 10$
    $\left(\frac{x}{7}\right)7 = 10 \cdot 7$
    $x = 70$

12. $3x(2x + 4y) + 7y(x - y) =$
    $6x^2 + 12xy + 7xy - 7y^2 =$
    $6x^2 + 19xy - 7y^2$

13. $-x(7x - 3y) + 6x(-y + 5x) =$
    $-7x^2 + 3xy - 6xy + 30x^2 =$
    $23x^2 - 3xy$

14. $-3(a + b) + a(-2 + b) =$
    $-3a - 3b + -2a + ab$
    $-3a - 3b - 2a + ab$
    $-5a - 3b + ab$

ANSWERS: FINAL TEST Page 2

15. Draw 3 tables and find 3 coordinates from these linear equations:

    $y = x - 4$

x	y
1	-3
2	-2
3	-1

    $y = 2x + 1$

x	y
1	3
2	5
3	7

    $y = 3x - 1$

x	y
1	2
2	5
3	8

16. Draw a graph and plot 3 points using this linear equation: $y = 2x - 1$.

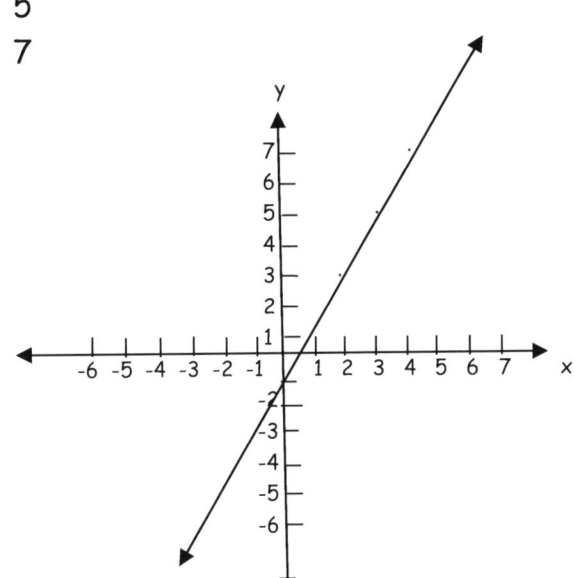

x	y
2	3
3	5
4	7

17. Look at the graph you drew for problem 16. Connect the dots and find the slope of that line by using the slopes formula.

$$m = \frac{y_2 - y_1}{x_2 - x_1}$$

$$m = \frac{5_2 - 3_1}{3_2 - 2_1} = \frac{2}{1} = 2$$

201

ANSWERS: FINAL TEST page 3

18. Find the y-intercept of the line from problem 17 by using the y-intercept formula:
$$y = mx + b$$
$$3 = 2 \cdot 2 + b$$
$$3 = 4 + b$$
$$3 - 4 = b$$
$$-1 = b$$

Look at the coordinates below and create a linear equation.

19. $(2,3)\ (3,4)$ $\quad m = \dfrac{4-3}{3-2} = \dfrac{1}{1} = 1$

$$3 = 1 \cdot 2 + b$$
$$3 = 2 + b$$
$$3 - 2 = b$$
$$b = 1$$

$$y = 1x + 1$$

20. $(2,1)\ (5,5)$ $\quad m = \dfrac{5-1}{5-2} = \dfrac{4}{3}$

$$1 = \dfrac{4}{3} \cdot 2 + b$$
$$1 = \dfrac{8}{3} + b$$
$$-\dfrac{8}{3} + 1 = b$$
$$-\dfrac{8}{3} + \dfrac{3}{3} = b$$
$$-\dfrac{5}{3} = b$$

$$y = \dfrac{4}{3}x - \dfrac{5}{3}$$

21. $(2,3)\ (4,1)$

$$m = \dfrac{1-3}{4-2} = \dfrac{-2}{2} = -1$$

ANSWERS: FINAL TEST page 4

$$3 = -1 \cdot 2 + b$$
$$3 = -2 + b$$
$$3 + 2 = b$$
$$5 = b$$

$$y = -1x + 5$$

22. What does it mean when the slope of a line is a negative number?
The line is going downhill on the graph.

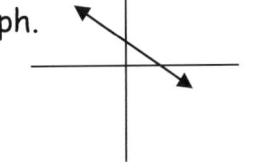

23. What does a linear equation make?

A linear equation makes a line.

24. What is the y-intercept?

The y-intercept is the point on the y-axis where the line crosses.

25. What does each letter in the formula below represent?

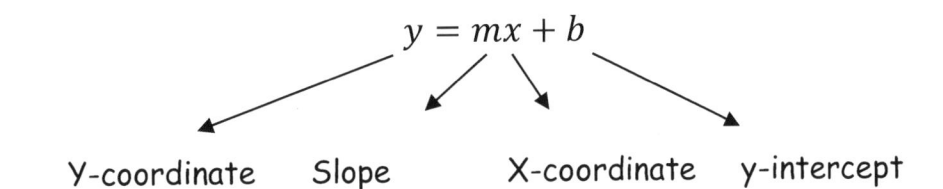

$$y = mx + b$$

Y-coordinate   Slope   X-coordinate   y-intercept

Made in the USA
Columbia, SC
11 August 2021